Are Electromagnetic Fields Making Me Ill?

Bradley J. Roth

Are Electromagnetic Fields Making Me Ill?

How Electricity and Magnetism Affect Our Health

 Springer

Bradley J. Roth
Department of Physics
Oakland University
Rochester, MI, USA

ISBN 978-3-030-98773-2   ISBN 978-3-030-98774-9   (eBook)
https://doi.org/10.1007/978-3-030-98774-9

This Springer imprint is published by the registered company Springer Nature Switzerland AG
The registered company address is: Gewerbestrasse 11, 6330 Cham, Switzerland

Acknowledgments

I thank the Kresge Library at Oakland University and the Rochester Hills Public Library for assisting me with obtaining books and articles related to this research. I appreciate comments on a draft of this book from Robert Bartholomew, Kenneth Foster, John Moulder, Auggie Roth, Luuk van Boekholdt, and Dilmini Wijesinghe.

Contents

About the Author

Bradley J. Roth obtained his undergraduate degree from the University of Kansas (1982) and a doctorate from Vanderbilt University (1987), both in physics. His graduate research was performed with John Wikswo, and resulted in the first detailed comparison of the transmembrane voltage and magnetic field produced by a single nerve axon.

In 1988, Roth joined the Biomedical Engineering and Instrumentation Program at the National Institutes of Health, where he worked on cardiac electrophysiology and magnetic stimulation of nerves. From 1995 to 1998, he was the Robert T. Lagemann Assistant Professor of Living State Physics at Vanderbilt. In 1998, he joined the Department of Physics at Oakland University, where he worked until his retirement in 2020.

Roth is coauthor with Russell Hobbie of the textbook *Intermediate Physics for Medicine and Biology*, and maintains a blog about the book at hobbieroth. blogspot.com. He was elected a fellow of the American Physical Society "for his theoretical and numerical studies of bioelectric and biomagnetic phenomena, especially for his contributions to the bidomain model of the heart."

Introduction

This book is about electric and magnetic fields, and their effect on your body. We will examine the use of magnets for pain relief, the risk of power line magnetic fields, the safety of cell phones, and the possibility that microwave weapons are responsible for the Havana syndrome. Many medical treatments are based on electromagnetism, including well established ones like heart pacemakers and neural prostheses, and more questionable ones such as bone healing, transcutaneous electrical nerve stimulation, and transcranial direct current stimulation. Innumerable books and articles have been written about each of these topics; my goal in this book is to examine them together, to get the big picture.

This book is *not* a research monograph. It presents no original discoveries and makes no attempt to be comprehensive. Moreover, it omits numerous details and technicalities that experts often argue about. It does, however, try to offer an overall view of the field that is accurate.

My target readers are nonscientists: journalists, politicians, teachers, students, and anyone who has heard about electric and magnetic fields interacting with biological tissue and wants to learn more. I use no mathematics, avoid jargon, and employ abbreviations only when repeating the same mouthful of words over and over again becomes tedious. I tried my best to make the book understandable to a wide audience.

One of my goals is to compare different phenomena and techniques. To do this, occasionally I need to be quantitative. Usually when I quote a number, it will be one of four quantities: an electric field, magnetic field, frequency, or specific absorption rate. I discuss the units of these quantities when each is introduced, but I will also give them to you now so you can watch for them later. The electric field is expressed in volts per meter (V/m), the magnetic field in tesla (T), the frequency in hertz (Hz), and the specific absorption rate in watts per kilogram (W/kg).

The numerical values of these quantities vary widely. Initially I did not plan to introduce metric prefixes or scientific notation, but that sometimes forces the reader to laboriously count zeroes to determine, for example, if the magnetic field is

© The Author(s), under exclusive license to Springer Nature Switzerland AG 2022
B. J. Roth, *Are Electromagnetic Fields Making Me Ill?*,
https://doi.org/10.1007/978-3-030-98774-9_1

0.00001 T or 0.000001 T. To avoid that, I shall make use of five common prefixes when I think they would be helpful. The list is given below.

Name	Symbol	Size	Common name
Micro	μ	0.000001	One millionth
Milli	m	0.001	One thousandth
Kilo	k	1000	One thousand
Mega	M	1,000,000	One million
Giga	G	1,000,000,000	One billion

For example, a millimeter (about the thickness of a dime) is one thousandth of a meter, and a radio wave with a frequency of one megahertz oscillates one million times per second.

Do not worry much about getting numbers exactly right. I don't care if you remember that your microwave oven uses a frequency of 2.45 GHz precisely, but only that you know your microwave operates at a few gigahertz. When we occasionally do a calculation, it will be a back-of-the-envelope estimate. I will not explain how to do intricate computations of electromagnetic fields. Instead, I want you to gain a general, intuitive appreciation of how electric and magnetic fields interact with the body. Perhaps you might be motivated to seek out a more specialized and technical explanation later, from some other source.

The book is organized by frequency. To start, we analyze static fields, next we consider fields changing at a leisurely 60 Hz, and then we examine fields varying at a rate that can excite our nerves and muscles. Finally, we discuss fields that oscillate millions, billions, or even trillions of times per second.

At the end of each chapter, I cite some references. These indicate where I obtained some of my information, and provide the source of any quotes. They also represent a bibliography that allows you to learn more about a topic. I could have cited thousands of papers, books, websites, and magazine articles, but I resisted the temptation.

Sometimes the effect of electric and magnetic fields is controversial. For any debate, I have tried to present both sides. Nevertheless, readers will soon catch on that I'm a skeptic. Each chapter title is a question, of which my answer is usually "probably not" or "no."

Now, let's begin.

Can Magnets Cure All Your Ills?

2

In their book *The Pain Relief Breakthrough* [1], Julian Whitaker and Brenda Adderly claim that magnetic fields dramatically reduce, and frequently eliminate, pain. They contend that placing magnets on your body cures backaches, arthritis, menstrual cramps, carpal tunnel syndrome, sports injuries, and more. These magnets are sometimes incorporated into bracelets or shoe insoles, or are strapped around wrists or knees. Selling such magnets is a multibillion-dollar business. Whitaker and Adderly tell the story of professional golfer Jim Colbert—known as "magnet man"— who revitalized his career by attaching magnets to his back. Apparently, magnets have become popular on the PGA tour.

Magnetism has always intrigued scientists and doctors. In the late eighteenth century, the German physician Franz Mesmer operated a magnetic clinic in Paris, where he treated ailments such as dropsy, gout, and scurvy (Fig. 2.1) [2]. King Louis XVI appointed a royal commission headed by Benjamin Franklin to investigate Mesmer's remedies, and it concluded that magnetic fields are ineffective [3]. Mesmer ended up fleeing Paris, and his so-called "animal magnetism" was debunked.

The Physics of Magnetism

Before we can understand how a magnetic field interacts with our bodies, we need to know what it is and how to measure it. The strength of a magnetic field is measured in tesla (T). The unit is named after Serbian-American engineer and inventor Nikola Tesla (Fig. 2.2), namesake of the Tesla electric car company and Thomas Edison's rival in creating our electrical power distribution system. An earlier unit of magnetic field is the gauss; 1 T equals 10,000 gauss. Much of the older literature expresses magnetic field strength in gauss, but in this book I shall always convert to tesla.

How big is a tesla? The earth produces a magnetic field of about 0.00005 T (50 µT, or half a gauss), and an ordinary refrigerator magnet has a field of about 0.001 T. A typical magnetic resonance imaging scanner uses a static magnetic field

© The Author(s), under exclusive license to Springer Nature Switzerland AG 2022
B. J. Roth, *Are Electromagnetic Fields Making Me Ill?*,
https://doi.org/10.1007/978-3-030-98774-9_2

Fig. 2.1 Patients in Paris receiving Franz Mesmer's animal magnetism therapy. Colored etching based on a painting by Claude Louis Desrais. (From the Wellcome Library No. 17918i, CC BY 4.0, https://creativecommons.org/licenses/by/4.0, via Wikimedia Commons)

Fig. 2.2 Nikola Tesla (1856–1943), age 34. The unit of magnetic field is named after him. (Napoleon Sarony, Public domain, via Wikimedia Commons)

of 1.5 T, although MRI units with stronger magnetic fields are becoming more common. The strength of the magnetic field used by Whitaker and Adderly for pain relief varies, but is usually about a tenth of a tesla (0.1 T).

Magnetic fields are produced by electric currents. These currents are either macroscopic, such as those flowing in a circuit, or microscopic, like those associated with spinning electrons. The magnetic field of the earth is created by macroscopic currents flowing in our planet's outer core (Fig. 2.3). An MRI scanner utilizes macroscopic currents circulating in superconducting coils that must be kept at ultracold temperatures. On the other hand, magnetic fields used for pain relief are produced by permanent magnets, in which electron spins line up to produce microscopic currents through a process called ferromagnetism.

Not only do currents produce magnetic fields, but conversely magnetic fields exert forces on currents. An electric motor works by passing current through a wire that sits in a magnetic field. The force acting on the wire causes it to move, thereby converting electrical energy in the wire into mechanical energy of motion. Sometimes the interaction of a magnetic field with a current is subtle. For instance, when you place two permanent magnets side-by-side, they repel each other because the magnetic field of one pushes on the microscopic current produced by electron spins in the other.

The magnetic field is a vector, meaning it has both strength and direction. Often the magnetic field is drawn as an arrow that points in the direction of the field and whose length indicates its strength. In a permanent magnet, the magnetic field vector points out of the north pole and into the south pole. Sometimes the vectors are connected into curves known as magnetic field lines. In that case, the direction of the magnetic field is parallel to the curve, and the strength is indicated by how closely the field lines are packed (Fig. 2.3).

The direction of the force that a magnetic field exerts on a current is complicated: it follows a "right-hand-rule" where the force is perpendicular to both the magnetic field direction and the direction that the current is flowing (Fig. 2.4). This complex relationship between the magnetic field, current, and force makes magnetism more difficult to understand than electricity.

Fig. 2.3 The magnetic field lines of the earth. The south magnetic pole is located near the north geographic pole. (Zureks, CC0, via Wikimedia Commons)

Fig. 2.4 The "right-hand
rule" relating the magnetic
field, current, and force

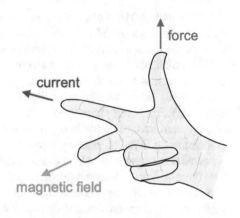

Biomagnetism

The term "biomagnetism" generally refers to the magnetic field produced by currents in the body [4]. The largest such field is made by the heart; its magnetic signal is called the magnetocardiogram, and it is sometimes used to diagnose heart disease. The magnetic field within the heart is about 0.00000001 T (or 0.01 μT). Usually the field is measured a short distance outside the chest, where its strength has dropped to 0.00001 μT. Many researchers are also interested in the magnetic field of the brain, referred to as the magnetoencephalogram. These fields are tinier still, about 0.000001 μT or smaller. Measuring the magnetoencephalogram provides a way to monitor the electrical activity of the brain without implanting any instrument in the body or even attaching electrodes to the scalp; merely place a magnetometer near the head and record the response. The magnetoencephalogram is so feeble—much less than one millionth of the earth's magnetic field—that often it must be measured with a special detector called a superconducting quantum interference device (SQUID), which needs to be kept at a frigid (cryogenic) temperature.

Occasionally researchers assert that biomagnetic fields can influence the body by altering nerve conduction [5–7]. This is not true. The magnetic field produced by a nerve is minuscule. When I was in graduate school, John Wikswo and I measured the magnetic field produced by a single giant nerve axon; it was only 0.0001 μT [8]. Such biomagnetic fields are so weak that they cannot possibly have any impact on neural behavior. Do not fall for the fallacy that since nerves can produce a magnetic field, that implies they can be influenced by a magnetic field.

Although most biological physicists would prefer to restrict the word "biomagnetism" to mean magnetic fields produced *by* the body, this term has been usurped by some to describe the influence of magnetic fields *on* the body. So-called "biomagnetic therapy" is similar to using magnets to treat pain. The technique was invented in 1988 by the Mexican surgeon Isaac Goiz Duran [9]. It uses two magnets (a "biomagnetic pair") that produce magnetic fields of about half a tesla, and are placed at specific locations on the body. Duran claims that these magnets act by influencing the pH of the blood (the blood's acidity), which affects the environment for pathogens such as viruses, bacteria, and other parasites. He believes that the body responds differently to "positive polarity" and "negative polarity" magnetic fields. Duran has trained many pupils at his Medical Biomagnetism School in Mexico City. He and his followers assert that biomagnetic therapy can cure a variety of diseases, including cancer, diabetes, AIDS, and Covid-19.

Some of Duran's ideas harken back to the bizarre notions of Albert Roy Davis and Walter Rawls, who believed that the north and south poles of magnets have dramatically different biological effects, even though the only difference between the poles is the direction of the field. For instance, they write that "when magnetic energy of the negative N [north] pole is applied to [a] cancer site, a remarkable reduction in the condition and also a marked arrest in further development of the cancer condition takes place… [whereas] when the S [south] pole of a magnet, this being the positive energy of a magnet, is applied to cancers they become more advanced and then develop, grow and spread at an accelerated rate" [10]. Their ideas are not limited to explaining how magnetic fields interact with biological tissue, but require a complete revision of the electromagnetic theory expressed in Maxwell's equations, which have formed the theoretical foundation for our understanding of electricity and magnetism for over 150 years, and are responsible for much of our modern technology (see Chap. 6). Indeed, Davis and Rawls immodestly declare that their readers "must be willing to leave behind them the outmoded, incorrect theories and concepts of magnetism [as formulated in Maxwell's equations]" and insist that their view "offers a totally different picture than is now used in present textbooks and is used as law and theory in all related research" [10].

Before examining the medical claims for magnetic and biomagnetic therapy put forth by Whitaker, Duran, Davis & Rawls, and others, we need to consider what evidence supports the idea that magnetic fields can have any biological effect at all. We begin by analyzing how the earth's magnetic field influences one of the smallest organisms: bacteria.

Magnetotactic Bacteria

The book *Iron, Nature's Universal Element* tells a charming tale about Richard Blakemore, a graduate student studying at the University of Massachusetts in Amherst during the mid 1970s [11]. His advisor gave Blakemore the task of analyzing bacteria in mud. As Blakemore peered through his microscope at these little organisms that were just 2 microns long (a "micron" is a nickname for a micrometer, one millionth of a meter, which is about one hundredth of the thickness of a human hair), he saw that many bacteria swam in a particular direction, and he jokingly started calling them "north-swimming" bacteria. A friend of his took the joke seriously, and placed a magnet next to the microscope. Suddenly, all the bacteria started swimming en masse in a different direction. Blakemore found he could change the direction they swam by moving the magnet. This was the first discovery of bacteria sensitive to a magnetic field: magnetotactic bacteria.

Fig. 2.5 A magnetotactic
bacterium. The orange dots
are the magnetosomes.
(CC BY-SA 3.0, https://
creativecommons.org/
licenses/by-sa/3.0, via
Wikimedia Commons)

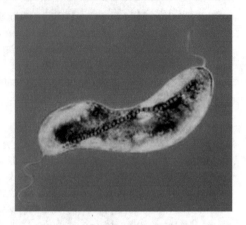

Blakemore and his colleagues imaged the bacteria's inner structure using an electron microscope. They found that inside a single bacterium are little particles, about one twentieth of a micron in diameter, wrapped in a membrane and made of magnetite, a form of iron oxide (Fe_3O_4). The most interesting behavior of these particles, called magnetosomes, is that in each bacterium about 20 of them were lined up like a string of pearls (Fig. 2.5).

Recall that a magnetic field can be created by spinning electrons. Most atoms have electrons that come in pairs, with one of the pair spinning opposite to the other. In that case their magnetic fields cancel out, and to a first approximation the material has no magnetic properties. A few elements, such as iron, are unusual in that their electrons do not all pair off. When iron atoms are packed close together their unpaired electrons all tend to spin in the same direction, creating a strong magnetic field. Magnetite has this property, as do refrigerator magnets and the magnets employed by Whitaker and Duran to treat patients.

The properties of magnetite are complicated, because magnetic materials often divide up into domains. Within a domain, several million unpaired electrons all line up, creating a magnetic field. A nearby domain will also have its electrons aligned, but in a different direction. The domains are less than a tenth of a micron in size, so a large piece of material contains millions of domains, which can all cancel out one another's magnetic field (unless an external magnetic field has "magnetized" or aligned them). A magnetosome, however, is the size of a single magnetic domain [11]. Each magnetosome acts like a current loop formed by its aligned spinning electrons. In other words, each magnetosome behaves like a little bar magnet.

An external magnetic field exerts a force on a bar magnet, tending to rotate it into alignment, and this force produces magnetic energy. If you calculate the magnetic energy associated with the interaction between a magnetosome and the earth's magnetic field, it is about 0.02 electron volts. (An electron volt is a unit of energy that is useful when discussing individual atoms and molecules.) We can compare this magnetic energy to the thermal energy. At room temperature, each molecule is jostling around in thermal motion, bumping into other objects around it and setting them in motion too. In 1827, English botanist Robert Brown observed the stirring and shaking of pollen grains colliding with water molecules, a phenomenon now known as Brownian motion. At room temperature, the kinetic energy associated with the thermal agitation of a molecule is about 0.02 electron volts. So, the magnetic energy of a magnetosome in the earth's magnetic field is roughly the same size as the thermal energy of molecules. If the magnetosome were smaller, its magnetic energy would be overwhelmed by the thermal energy, making any biological response unlikely. The bacteria's detection of the earth's magnetic field is more reliable if the magnetic energy is greater than the thermal energy, so magnetotactic bacteria line up many magnetosomes, which together act like a tiny compass.

The story of magnetotactic bacteria teaches us two important lessons.

1. The easiest way for an organism to sense a magnetic field is for it to contain iron. Not just a few isolated iron atoms—as you find in the blood protein hemoglobin—but lots of densely-packed iron. You need magnetite, or some other mineral similar to it. If magnetite is present, magnetic effects on an organism are plausible.
2. The interaction of a magnetic field can be measured by its magnetic energy, which can be compared to the thermal energy. If the thermal energy dominates, the magnetic response is swamped by the random tumbling of molecules. If the magnetic energy dominates, then biological effects are possible. We will reconsider such comparisons with thermal motion later in this book; it is a key concept.

Magnetoreception

Bacteria are not the only organisms able to sense the earth's magnetic field; so can birds, bees, and butterflies [11]. In 1963, German zoologist Wolfgang Wiltschko placed European robins inside a chamber and turned on a magnetic field comparable in strength to the earth's field. He did not expect a response, but to his surprise the birds oriented with the field. This experiment was performed before Blakemore's discovery of magnetosomes in bacteria, so no one understood how the robins could possibly detect the field. Yet, the robins proved adept at sensing magnetic signals during their annual migration. Wiltschko supposed that the birds would behave similar to hikers, using a compass to orient in the proper direction, and would respond to the magnetic field's horizontal component. Instead, these experiments showed that robins detect the field's vertical component—they use a "dip compass." The vertical component provides information about latitude: the distance north or south of the equator.

Another bird capable of sensing a magnetic field is the homing pigeon. Magnetite has been found in the beak of these pigeons, and it presumably acts as the mechanism for sensing the earth's magnetic field, an ability called magnetoreception. In addition, other animals contain magnetite and are able to detect magnetic fields, such as rainbow trout, sockeye salmon, and sea turtles [11]. Magnetoreception acts as a sixth sense that these animals use for navigation. However, the magnetic detection process is still not completely understood.

Most sense organs have been optimized by evolution over millions of years. As a result, they are sensitive enough to respond to minute signals. For instance, our ears can detect sounds so soft as to be just above the thermal limit; if they were much more responsive we would hear the constant roar of Brownian motion crashing against our eardrums. Our eyes can see a single photon of light. Just because an animal's exquisitely sensitive magnetoreceptor can sense the earth's magnetic field does not mean that the earth's field has a profound influence on that animal's physiology. It only means that some animals are able to detect minuscule magnetic fields.

Can people detect magnetic fields? Caltech geophysicist Joseph Kirschvink has found magnetite in the brain, which could be the basis of magnetoreception in humans [12]. Experiments to test this hypothesis are difficult; contamination of tissue samples is always a problem, and the mere presence of magnetite does not by itself imply that a magnetic sensor exists. Recently, Kirschvink built a shielded room in which he varies the magnetic field direction while keeping its strength approximately equal to that of the earth [13]. He claims to have detected a change in the electric signal from the brain (the electroencephalogram) following rotation of the magnetic field vector, implying a subconscious perception of the magnetic field. The evidence for human magnetoreception is tantalizing but inconclusive. It's too soon to say for sure. I have my doubts.

Recently, another mechanism for magnetic field detection, not involving magnetite, has been proposed. In general, magnetic fields do not influence chemical reactions [14]. To appreciate why, consider a reaction that involves a single electron. The magnetic energy of an electron interacting with the earth's magnetic field is a few billionths of an electron volt. The thermal energy of an electron at room temperature is a few hundredths of an electron volt. The thermal energy is more than a million times greater than the magnetic energy. Unless you have millions of electrons all acting together, as in a magnetosome, any magnetic field effect is lost in the noise. Therefore, simple chemical reactions involving a handful of electrons—such as those leading to, say, changes in the pH of water—are unaffected by the magnetic field strength.

An exception to the rule that chemical reactions do not depend on the magnetic field strength involves free radicals: molecules that have unpaired electrons in their outer, or valence, shell [15]. (Iron contains unpaired electrons in a deep inner shell, and is not a free radical.) An example is the hydroxyl radical, OH, where the outer shell of the oxygen atom has a single unpaired electron (do not confuse the hydroxyl radical with the hydroxide ion, OH^-, which has no unpaired electrons but does carry a negative charge). Free radicals are reactive, have short lifetimes, and are rarely found in our cells at high concentrations. However, some reactions between pairs of free radicals are unusually sensitive to magnetic fields. In the lingo of quantum mechanics, the electrons in these free radicals undergo a singlet-to-triplet conversion that is magnetic field dependent. A few animals, including the European robin, may take advantage of free radical reactions instead of magnetite for magnetoreception. Sönke Johnsen and Kenneth Lohmann, after reviewing the data, conclude that "the current evidence for the radical-pair hypothesis is maddeningly circumstantial" [15]. The jury is still out on this issue.

Magnetic Resonance Imaging

Magnetotactic bacteria and magnetoreception demonstrate that some organisms are able to sense a magnetic field. Can a magnetic field alter human physiology? To answer this question, we should examine how our bodies respond to strong magnetic fields. The strongest field most people ever experience is during magnetic resonance imaging (Fig. 2.6).

A common MRI scanner immerses the patient in a 1.5 T magnetic field, which is 30,000 times stronger than the earth's field. Scanners with more intense fields, 4 T and even larger, are becoming more common. MRI has revolutionized medical diagnosis, providing superb images of soft tissue and allowing a glimpse into the brain. But is MRI safe? An advantage of MRI over other types of imaging, such as computed tomography (CT or "CAT scan"), is that the patient is not exposed to any x-rays. On the other hand, tomography does not expose the patient to a strong magnetic field. What are the safety hazards associated with the magnetic field of a magnetic resonance imaging scanner?

The biggest hazard related to MRI is when ferromagnetic objects, usually those containing iron, are sucked into the scanner by accident. The magnetic field interacts with all the iron's spinning electrons to pull them into the region of stronger field. For instance, if a screwdriver is set down near the scanner, it might be yanked into the powerful magnetic field and turn into an arrow that could pierce the machine or, heaven forbid, skewer the patient. Even worse, think what an iron gas cylinder would be like as a torpedo!

Fig. 2.6 Patient being positioned for an MRI study of the head and abdomen. The scanner is a Siemens MAGNETOM Aera (1.5 T superconducting magnet). (Ptrump16, CC BY-SA 4.0, https:// creativecommons.org/licenses/by-sa/4.0, via Wikimedia Commons)

Apart from being hit by a flying object, are there any other risks associated with a strong, static (constant in time) magnetic field? Any risk must be slight, because millions of MRIs have been performed with no ill effects. John Schenck—a scientist at General Electric whose brain was the first to be imaged using strong-field MRI—has reviewed all potential mechanisms by which magnetic fields interact with our bodies, including forces on moving objects, changes to blood flow, and effects of the weak magnetic properties of tissues [16].

A possible hazard arises when a patient moves while in an intense magnetic field. We noted earlier that magnetic fields exert a force on a current. But an electric current is simply moving charge. Much of our body consists of salt water, with the sodium, potassium, and chloride ions that are dissolved in our bodily fluids being the most abundant free charges. When a patient lies motionless in a MRI scanner, the magnetic field exerts no force on these stationary charges. But if the patient moves around in the field, all those sodium, potassium, and chloride ions move too. And since they are moving in a magnetic field, they experience a force. The sodium and potassium ions carry a positive charge, while the chloride ions carry a negative charge. Since their magnetic forces are in opposite directions, the charges tend to separate.

Regions of positive and negative charge create an electric field, which is the main topic of Chap. 3. For typical speeds and magnetic field strengths during MRI, the motion-induced electric fields are small. However, our body has evolved sensitive sensory organs that may be triggered by these small electric fields. In 1896, French scientist Jacques-Arsène d'Arsonval reported seeing flashes of light associated with a magnetic field (magnetophosphenes). Patients undergoing a MRI scan have reported seeing magnetophosphenes, perceiving a metallic taste in their mouth, or experiencing vertigo from activation of their inner ear. All these symptoms are mild, result in no permanent damage, and are more prevalent in stronger magnetic fields, such as those found in the more modern 4 or 7 T MRI scanners. Because they are associated with motion, these symptoms are more likely to occur when you enter or exit a scanner, and not when you are lying quietly in one. There is a push to build scanners with even more powerful magnetic fields; 11 T MRI devices have been constructed for research purposes, and higher strengths are being considered. Schenck believes that motion-induced electric fields and their resulting side effects could ultimately limit how strong MRI magnetic fields can be.

Another potential safety hazard caused by a strong, static magnetic field is changes to blood flow. Even if you are lying still during a MRI scan, your blood is moving. This motion could induce small electric fields across a blood vessel. The speed of blood flow is greatest in large vessels, so the most likely place to experience any change of blood flow is in big arteries, such as the aorta. Magnetic forces act on moving ions in the blood and then friction between the ions and water transmits this force to the blood as a whole, potentially influencing the flow. This effect, called magnetohydrodynamics, is negligible in typical MRI scanners, but could potentially become significant in high magnetic field devices [17].

Yet another possible interaction mechanism arises from the magnetic properties of tissue. Usually spinning electrons cannot line up due to their random thermal motion. However, a strong magnetic field results in a tiny fraction of the spins becoming aligned with the field, a process called paramagnetism. The size of this response is measured by a dimensionless number known as the magnetic susceptibility. In most biological tissue, the magnitude of the susceptibility is on the order of 0.0001. This small value means that biological tissues have almost no influence on magnetic fields; they pass through the tissue almost as if it were not there. Nevertheless, when different tissues have different susceptibilities, they may experience weak magnetic forces. Like a screwdriver or oxygen cylinder, a paramagnetic tissue is pulled into a region of stronger magnetic field. For tissue, however, the force is usually negligible.

An example of a force arising from magnetic susceptibility is found in blood. Red blood cells contain iron, but the few iron atoms are far apart, encased within the hemoglobin protein. The distant iron atoms cannot cooperate, as they do in magnetite, to produce a big magnetic response. They do, however, cause blood to have a different magnetic susceptibility than the surrounding tissue; it is paramagnetic. Schenck estimates that magnetic forces on red blood cells are slight; less than 4% of the gravitational force on these cells, and gravitational forces are so weak that they produce few physiological changes to blood flow (for instance, gravity does not cause red blood cells to settle to the bottom of an artery). Susceptibility effects grow with the square of the magnetic field strength. This is because the magnetic field does two things: align the spins and exert a force on them. So if the magnetic field in an MRI scanner increases by a factor of 10 (say, from 1.5 to 15 T), forces arising from susceptibility differences increase by a factor of 100. These forces become increasingly important at higher magnetic field strengths.

Water exhibits a weak magnetic behavior termed diamagnetism. A diamagnetic material is not attracted into a magnetic field, but instead is repelled by it. The susceptibility is small, so any forces during an MRI scan are inconsequential. In a remarkable experiment, however, Andrey Geim—a Nobel Prize-winning physicist from the University of Manchester—used a 16 T magnetic field to levitate a frog [18, 19]. Frogs are mostly water, so their diamagnetic tissue creates a magnetic force that opposes the gravitational force (this trick works only for diamagnetic materials, not paramagnetic or ferromagnetic). To levitate, you need both a powerful magnetic field and a big field gradient: the field must vary dramatically over a short distance. The frog did not appear to suffer any side effects from its experience defying gravity.

Finally, we learned earlier that magnetic fields rarely affect the rate of a chemical reaction. Patients are exposed to magnetic fields in a MRI scanner that are significantly stronger than those present during Duran's biomagnetic therapy, but they experience no chemical alterations to their blood, such as a change in pH.

Schenck's bottom line is that the large static magnetic field of an MRI scanner has almost no influence on the human body. The main influence is due to motion in the magnetic field, but for current MRI scanners this effect is small and transient. Some potential problems could emerge in proposed ultra-high-field scanners, but even these probably generate symptoms that are mild. The excellent safety record for magnetic resonance imaging gives us confidence that static magnetic fields are safe.

Critical Reviews

Studying how magnetic fields affect biology can be frustrating. You can find peer-reviewed scientific studies supporting all points of view. Hundreds of papers of varying quality have been published describing laboratory experiments, and generally dozens of these reports are consistent with your opinion regardless of what your opinion is. I believe this is the primary reason why the topic of biological responses to magnetic fields remains controversial.

Often a critical review helps sort out the good studies from the bad. In 1999, Leonard Finegold, an emeritus professor of physics at Drexel University, published such a review in the *Scientific Review of Alternative Medicine* [20]. He analyzed several articles that had reported biological consequences of static magnetic fields, and identified weaknesses in the experiments. For instance, in one study the connective tissues in the joints of a rat were damaged. The authors claimed magnetic fields reduced inflammation in these rats, but Finegold noted that if the rats move around in the field they might sense motion-induced electric fields and therefore choose to move less. In that case, the rats differ not only by their exposure to a magnetic field but also by how much they move, and we cannot tell if the observed effect on inflammation is caused by the magnetic field or by the increased movement.

In another example, sometimes a big permanent magnet is placed near an animal to test if a magnetic field has any influence on its behavior. As a control, the experiment is repeated but the magnet is removed so no field is present. However, if the presence of a massive magnet affects the temperature, humidity, air circulation, exposure to light, or some other factor, the experimenter might not be able to tell if the magnetic field itself is causing the effect. For the reviewer, sorting out such issues requires carefully reading each article's methods section (was the magnet replaced by a similar big lump of metal that is not magnetic?). This is why scientific papers contain a detailed description of how an experiment was performed. After analyzing dozens of studies, Finegold concluded that magnetic fields have little or no effect on our physiology [20, 21].

Clinical Trials

So far, much of our discussion has focused on mechanisms: how does a response happen. This mechanistic approach is usually taken by physicists and engineers. But most medical doctors do not care *how* magnetic fields interact with the body, but only *if* magnetic fields interact with it. In particular, they are concerned mainly about a magnet's usefulness in the clinic. For example, do static magnetic fields alleviate pain? To answer these types of questions, we need clinical trials.

The problem is, many clinical trials have been conducted to test the hypothesis that magnetic fields reduce pain, and the results are not consistent. What we require is a procedure to weigh the pros and cons of all the clinical trials, placing greatest weight on the best ones. What we need is "meta-analysis" [22].

Meta-analysis is a formal, quantitative method of systematically assessing a group of clinical trials, to reach conclusions that are more reliable than those of any individual study. In 2007, Max Pittler, Elizabeth Brown and Edzard Ernst published "Static Magnets for Reducing Pain: Systematic Review and Meta-Analysis of Randomized Trials" [23]. They analyzed 29 relevant clinical trials. The best of them were placebo-controlled, randomized, and double-blind.

All the trials were placebo-controlled. This means that some patients received treatment with real magnets, and others were treated with objects that resembled magnets but produced a much weaker magnetic field or no magnetic field at all (the placebo). Doctors know that sometimes a patient responds positively if they *think* they are being treated, regardless of whether or not they receive the treatment. This is a problem, especially for studies that assess subjective judgement, such as the severity of pain. In a placebo-controlled experiment, the patient cannot determine if they are receiving a treatment or merely a well-designed placebo. Whenever possible, clinical trials should be placebo-controlled.

All the trials were randomized. That means that patients were placed in the treatment or placebo groups at random, which avoids spurious correlations. For example, you do not want patients in the placebo group to consist of more men, while the treatment group is more women, since then you would have difficulty distinguishing treatment effects from gender effects. Similarly, you do not want the treatment group to consist of young patients with the placebo group having old patients, or poor patients compared to rich patients, or fat patients compared to thin patients. By assigning patients to the placebo or treatment group at random, you expect on average equal numbers of men and women in each group, equal numbers of young and old, rich and poor, fat and thin. This averaging works best in large studies, where random fluctuations are less likely.

Finally, most of the trials were double-blind. Obviously, you do not want patients to know if they are in the placebo or treatment group, otherwise the placebo effect may not work its magic. That is, you want a blind study, with patients not aware whether or not they received the actual treatment. Double-blinding means that not only the patient, but also the physician, does not know who is in the placebo or treatment group. This eliminates any subtle hints that a physician might let slip when delivering the treatment. Try as they might to treat each patient the same, it is difficult for doctors to avoid providing inadvertent clues if they know which group a patient is in.

The finest clinical trials are therefore placebo-controlled, randomized, double-blind, and have a large number of patients in the study to avoid statistical fluctuations. A meta-analysis takes all these factors, and others, into account when coming to conclusions.

In their meta-analysis, after weighing the evidence from all the clinical trials, what did Pittler, Brown, and Ernst conclude? They wrote "The evidence does not support the use of static magnets for pain relief, and therefore magnets cannot be recommended as an effective treatment. For osteoarthritis, the evidence is insufficient to exclude a clinically important benefit, which creates an opportunity for further investigation" [23]. In other words, in most cases the evidence goes against magnets providing pain relief. In the case of arthritis, there is not enough evidence to say one way or the other.

Point/Counterpoint

Each issue of the journal *Medical Physics*, published by the American Association of Physicists in Medicine, contains a point/counterpoint article. In these articles, an idea is proposed and then two prominent researchers debate it, one for and one against. These articles are excellent teaching tools, as they illustrate how scientists deal with controversial topics. Anyone who grew up in the 1970s probably remembers the point/counterpoint debates on the television show *60 Minutes*, in which James Kilpatrick and Shana Alexander would wrangle over a political issue (Fig. 2.7). They also might recall the hilarious spoof of these quarrels by Dan Aykroyd and Jane Curtin on *Saturday Night Live*. Point/counterpoint articles are medicine's version of these *60 Minutes* debates.

In the July 2008 issue of *Medical Physics*, the proposition of the point/counterpoint article was "despite widespread use there is no convincing evidence that static magnets are effective for the relief of pain" [24]. Arguing for the proposition was Max Pittler, the lead author of the meta-analysis we recently discussed [23]. Arguing against it was Tim Harlow, a medical doctor who treats patients using complementary medicine [25]. Pittler led off by summarizing his previous analysis. Harlow's opening statement centered on two points. First, he noted that even though we do not know how something works, it might work nevertheless. He writes "to consider something impossible simply because we do not yet understand it is hubris" [24]. Harlow's second point deals with the placebo effect. He notes that, regardless of how they work, magnets help many of his patients deal with pain. "Even if it were entirely a placebo effect, static magnets would still be an effective treatment for the relief of pain and certainly would be far safer than conventional drugs" [24].

My favorite part of a point/counterpoint article is the rebuttal. Pittler countered Harlow by saying "promoting placebo treatment is counterproductive. It is expensive; costs the patients and, in some countries, the taxpayer; undermines rational

Fig. 2.7 Point/Counterpoint debate during the television show *60 Minutes*. (With permission, from itsabouttv.com)

thinking; and opens the door to a plethora of quack treatments" [24]. I agree, but an even more important consideration is trust. Patients ought to trust their doctors. If doctors prescribe treatments that they know do not work—but may benefit the patient by the placebo effect—that means these physicians must lie to their patients. Such dishonesty undermines the patient-doctor relationship and weakens trust in the entire medical profession. I want my doctor to tell me the truth, even if that means I do without the benefits of the placebo effect. Would I still feel this way if I were experiencing agonizing, chronic pain? I don't know.

Harlow responds to Pittler by asserting that "meta-analysis does not prove beyond question that magnets do not relieve pain, only that the evidence for relief of pain is not 'convincing.' Patients who experience pain relief with magnets would not be so convinced" [24]. With any alternative medical treatment, anecdotal evidence may be persuasive to many people. I am not suggesting that anecdotal evidence is of no value, but merely that for almost any illness, some people get better and some do not. If you give patients just about any treatment, no matter how silly, some of them are bound to improve. You can then point to those people and tell their story to justify the effectiveness of the treatment. The main purpose of randomized, placebo-controlled, double-blind clinical trials is to go beyond anecdotes and determine if a treatment has a real effect.

When I taught medical physics to college students, we spent 20 minutes each Friday afternoon discussing a point/counterpoint article. One feature of these articles that makes them such an outstanding teaching tool is that there exists no right answer, only weak or strong arguments. Science does not proceed by proclaiming universal truths, but by accumulating evidence that allows us to be more or less confident in our hypotheses. Conclusions beginning with "the evidence suggests…" are the best science has to offer.

Another Clinical Study

In 2007, another clinical study by Soledad Cepeda and her colleagues was published in the journal *Anesthesia & Analgesia* [26]. It appeared too late to be included in Pittler et al.'s meta-analysis, but supported their conclusion. Cepeda et al.'s study was placebo-controlled, randomized, and double-blind. It differed from previous studies in that it examined acute post-operative pain rather than chronic pain. One problem with chronic pain studies is that patients might, given enough time, figure out if they have a real magnet or a placebo that looks like a magnet but is not magnetic. Think about it; if you were a patient in the study, when you went home and were in the privacy of your bedroom, could you resist placing a paper clip next to your "magnet" to see if it was attracted? Placebos should be constructed carefully to resemble the therapeutic device in every way possible, or they may not do their job correctly.

A disadvantage of the study by Cepeda and her coworkers is that it could not look at pain relief that builds up over time. Many promoters of magnet therapy, however, insist their therapeutic response occurs quickly. Cepeda et al.'s verdict was unambiguous: "Magnetic therapy lacks efficacy in controlling acute postoperative pain intensity levels or opioid requirements and should not be recommended for pain relief in this setting" [26].

Cepeda et al.'s article describing their clinical trial was accompanied by an editorial by Bruce Flamm, a clinical professor of obstetrics and gynecology at the University of California, Irvine. Flamm revealed his skeptical view in the editorial's title: "Magnetic Therapy: Healing or Hogwash?" [27] Editorials are often more frank than scientific papers. Flamm wrote "as documented by Cepeda et al., it is crystal clear that billions of dollars have already been spent on magnet therapy, or perhaps, *wasted* on magnet therapy. To be blunt, there is no proven benefit to magnet therapy" [27].

Conclusion

The *British Medical Journal* is one of the premier publications in all of medicine. In 2006, Finegold and Flamm teamed up to publish an editorial titled "Magnet Therapy: Extraordinary Claims, But No Proved Benefits." After reviewing the evidence for and against magnet therapy, their final paragraph concluded

> Extraordinary claims demand extraordinary evidence. If there is any healing effect of magnets, it is apparently small since published research, both theoretical and experimental, is weighted heavily against any therapeutic benefit. Patients should be advised that magnet therapy has no proved benefits. If they insist on using a magnetic device they could be advised to buy the cheapest—this will at least alleviate the pain in their wallet.[28]

Physicist Robert Park took an even more negative view of magnet therapy. In his book *Voodoo Science*, he compared the biological effects of magnetic fields to the alternative medicine treatment of homeopathy, in which a drug is repeatedly diluted until not a single molecule of the active ingredient remains [29]. Park wrote "Not only are magnetic fields of no value in healing, you might characterize these as 'homeopathic' magnetic fields" [30].

The conclusions of this chapter are bad news for those such as Julian Whitaker and Isaac Goiz Duran who peddle magnets for pain relief. The treatment does not work. There is, however, some good news for those embracing magnet therapy: it has few side effects and is generally safe. If magnets do nothing, then they do nothing bad.

References

1. Whitaker, J., & Adderly, B. (1998). *The pain relief breakthrough: The power of magnets to relieve backaches, arthritis, menstrual cramps, carpal tunnel syndrome, sports injuries, and more*. Little Brown and Co.
2. Fara, P. (2005). *Fatal attraction: Magnetic mysteries of the enlightenment*. Icon Books.
3. Gould, S. J. (1989). The chain of reason vs. the chain of thumbs. *Natural History, 98*, 12–21.
4. Hobbie, R. K., & Roth, B. J. (2015). *Intermediate physics for medicine and biology* (5th ed.). Springer.

5. Isakovic, J., Dobbs-Dixon, I., Chaudhury, D., & Mitrecic, D. (2018). Modeling of inhomogeneous electromagnetic fields in the nervous system: A novel paradigm in understanding cell interactions, disease etiology and therapy. *Scientific Reports, 8*, 12909.

6. Melendy, R. F. (2018). A single differential equation description of membrane properties underlying the action potential and the axon electric field. *Journal of Electrical Bioimpedance, 9*, 106–114.

7. Wang, H., Wang, J., Cai, G., Liu, Y., Qu, Y., & Wu, T. (2021). A physical perspective to the inductive function of myelin—A missing piece of neuroscience. *Frontiers in Neural Circuits, 14*, 562005.

8. Roth, B. J., & Wikswo, J. P. (1985). The magnetic field of a single axon: A comparison of theory and experiment. *Biophysical Journal, 48*, 93–109.

9. https://escuelaisaacgoiz.com. Access date: January 12, 2022.

10. Davis, A. R., & Rawls, W. C., Jr. (1974). *Magnetism and its effects on the living system.* Exposition Press.

11. Mielczarek, E. V., & McGrayne, S. B. (2000). *Iron, nature's universal element: Why people need iron & animals make magnets.* Rutgers University Press.

12. Kirschvink, J. L., Kobayashi-Kirschvink, A., & Woodford, B. J. (1992). Magnetite biomineralization in the human brain. *Proceedings of the National Academy of Sciences, 89*, 7683–7687.

13. Wang, C. X., Hilburn, I. A., Wu, D.-A., Mizuhara, Y., Cousté, C. P., Abrahams, J. N. H., Bernstein, S. E., Matani, A., Shimojo, S., & Kirschvink, J. L. (2019). Transduction of the geomagnetic field as evidenced from alpha-band activity in the human brain. *eNeuro, 6*, ENEURO.0483-18.2019.

14. Atkins, P. (1976). Magnetic field effects. *Chemistry in Britain, 12*, 214–218.

15. Johnsen, S., & Lohmann, K. J. (2008). Magnetoreception in animals. *Physics Today, 61*, 29–35.

16. Schenck, J. F. (2005). Physical interactions of static magnetic fields with living tissues. *Progress in Biophysics & Molecular Biology, 87*, 185–204.

17. Tenforde, T. S. (2005). Magnetically induced electric fields and currents in the circulatory system. *Progress in Biophysics & Molecular Biology, 87*, 279–288.

18. Geim, A. (1998). Everyone's magnetism. *Physics Today, 51*, 36–39.

19. https://www.youtube.com/watch?v=A1vyB-O5i6E. Access date: January 12, 2020.

20. Finegold, L. (1999). The physics of "alternative medicine": Magnet therapy. *Scientific Review of Alternative Medicine, 3*, 26–33.

21. Finegold, L. (2012). Resource letter BSSMF-1: Biological sensing of static magnetic fields. *American Journal of Physics, 80*, 851–861.

22. Haidich, A. B. (2010). Meta-analysis in medical research. *Hippokratia, 14*(Suppl 1), 29–37.

23. Pittler, M. H., Brown, E. M., & Ernst, E. (2007). Static magnets for reducing pain: Systematic review and meta-analysis of randomized trials. *Canadian Medical Association Journal, 177*, 736–742.

24. Pittler, M. H., Harlow, T., & Orton, C. G. (2008). Despite widespread use there is no convincing evidence that static magnets are effective for the relief of pain. *Medical Physics, 35*, 3017–3019.

25. Harlow, T. (2021). *That something else: A reflection on medicine and humanity.* Austin Macauley.

26. Cepeda, M. S., Carr, D. B., Sarquis, T., Miranda, N., Garcia, R. J., & Zarate, C. (2007). Static magnetic therapy does not decrease pain or opioid requirements: A randomized double-blind trial. *Anesthesia & Analgesia, 104*, 290–294.

27. Flamm, B. L. (2007). Magnet therapy: Healing or hogwash? *Anesthesia & Analgesia, 104*, 249–250.

28. Finegold, L., & Flamm, B. L. (2006). Magnet therapy: Extraordinary claims, but no proved benefits. *British Medical Journal, 332*, 4.

29. Ernst, E. (2016). *Homeopathy: The undiluted facts.* Springer.

30. Park, R. L. (2000). *Voodoo science: The road from foolishness to fraud.* Oxford University Press.

Can a 9-Volt Battery Make You Smarter?

3

On the November 13, 2020 episode of the reality television show *Shark Tank* (Season 12, Episode 5), two earnest entrepreneurs, Ken and Allyson Davidov, try to persuade five hard-nosed investors, the "sharks," to buy into their company. Ken and Allyson sell LIFTiD, a device that applies a weak, steady electrical current to the head. Ken said it's supposed to improve "productivity, focus, and performance." Allyson claimed it's a "smarter way to get a… boost of energy." Ken called it "the coffee of the future" [1].

Each shark tried out the apparatus, which required attaching two saline-soaked electrodes to their forehead. When turned on, there were many oohs and aahs. Shark Mark Cuban exclaimed "I feel like I'm being shocked," shark Kevin O'Leary declared "I feel it in my upper teeth," and shark Lori Greiner said it was "freaking me out" [1]. Ken explained that LIFTiD is an emerging technology based on transcranial direct current stimulation. He claimed it works by neuroplasticity (changing the electrical structure of the brain) and is backed by 5000 published studies. According to Ken, LIFTiD stimulates the frontal lobe of the brain, and should be applied for 20 minutes a day while doing some activity such as reading a book, studying for an exam, or playing a video game. Ken and Allyson offered the sharks 10% of their company if they would invest $200,000.

In the previous chapter, we examined the biological consequences of static magnetic fields. LIFTiD does not use a *magnetic* field, but instead produces a static (DC, or direct current) *electric* field. In this chapter, we look at the biological effects of static electric fields. Before we evaluate LIFTiD and transcranial direct current stimulation in more detail, we must examine the underlying physics of electricity.

B. J. Roth, *Are Electromagnetic Fields Making Me Ill?*,
https://doi.org/10.1007/978-3-030-98774-9_3

Electrostatics

Electric fields are produced by charge, which comes in two types: positive and negative. In a device like LIFTiD, one electrode is negatively charged (the cathode), and the other is positively charged (the anode). The electric field points out of the anode and into the cathode. In addition, an electric field exerts a force on charge. When you put these two ideas together—charge produces an electric field, and an electric field produces a force on charge—you arrive at the well-known rule that like charges repel and unlike charges attract.

Like the magnetic field, the electric field is a vector, and has both a strength and a direction. Electrostatics (the study of static charges, electric fields, and their forces) is simpler than magnetostatics, because in electrostatics the electric field and the resulting force are both in the same direction. Also, the electric force does not require the charge to be moving, as was needed for the magnetic force. Stationary charges experience electric forces.

Electric fields are measured in volts per meter (V/m). The field required to excite a neuron, or nerve cell, in the brain is about 10 V/m. A stronger electric field of about 500 V/m can provoke a deadly heart arrhythmia. Inside your microwave oven the electric field strength is about 1000 V/m. In the next chapter we examine electric fields under power lines, whose strength is about 15,000 V/m. To find gigantic electric fields, look in the cell membrane, where you typically find a field of roughly 10,000,000 V/m.

In this book, we frequently discuss the biological influence of weak electric fields. If you ever had an electrocardiogram, your doctor was measuring the electric field generated in your torso by your heart. These fields have a strength of about 0.002 V/m. A shark has a specialized sensory organ that lets it detect an electric field as small as 0.000001 V/m, or equivalently 1 μV/m.

How is the electric field, in volts per meter, related to the voltage, in volts? Voltage is a concept associated with energy. If you push a positive ion against an electric field, you give it electrical energy, just as you give gravitational energy to a boulder that you roll uphill. The amount of energy per unit charge is the voltage, just as the electrical force per unit charge is the electric field. You can calculate the voltage as the electric field multiplied by the distance the charge moves. For instance, if you push a sodium ion against the electric field of about 10,000,000 V/m (10 MV/m) in a cell membrane, which has a thickness of about 0.00000001 meters (0.01 microns), you give it a voltage of a tenth of a volt, or 100 millivolts. This is roughly the size of the resting voltage across a typical nerve membrane. In this book, we talk mainly about electric fields in volts per meter, but refer occasionally to the voltage in volts.

The salt water of our body contains charged ions such as sodium, potassium, and chloride. If an electric field exists in our body it causes these ions to move, producing an electric current. Most medical papers prefer to report the current (in amps, which is charge per second) or the current density (current per unit area) instead of the electric field. However, the electric field is a more fundamental quantity in electromagnetism than current or current density, so in this book I always express current in terms of its associated electric field. To convert from one to the other, note that the current density is equal to the electric field times the conductivity, where conductivity measures how easily charge moves in a material. Salt water is a conductor; it has a relatively high conductivity compared to an insulator such as rubber or air, although it has a relatively low conductivity compared to a metal like copper. Because biological tissue is composed mostly of salt water, tissue also is a conductor.

If a conductor, surrounded by air, is placed in an electric field, that field exerts a force on the charges in the conductor, causing them to move. Where do they go? Unless the electric field is huge, the charges cannot leave the conductor and enter the insulting air, so they pile up on its surface. The resulting surface charge creates its own electric field pointing opposite to the original field. This process continues until the external field applied to the conductor and the internal field produced by surface charge cancel; their sum, the total electric field in the conductor, is zero. A person's body is a conductor. Therefore, when people are exposed to an electric field, charge accumulates on their skin but no electric field exists anywhere inside their body; their tissues are shielded from an external field (Fig. 3.1). Such shielding is one reason the influence of static electric fields on our body is so small: the electric field cannot penetrate into our body and affect our cells. This behavior is in stark contrast to a static magnetic field, which passes through our body unimpeded.

As a specific example, there is a static electric field in the earth's atmosphere that is pointing downward and has a strength of about 100 V/m [2]. We live our lives in this field. But inside our bodies no field is present, because we are a conductor and therefore are shielded from it. If this field could penetrate our bodies, it would be strong enough to activate our nerves and excite our heart.

If our body is shielded from an electric field, how does a device like LIFTiD work? The answer is that LIFTiD uses electrodes. The electric field created by LIFTiD causes charged ions in our body to move, and these charges tend to

Fig. 3.1 When a spherical conductor is placed in an otherwise uniform electric field (blue curves), positive (red) and negative (green) charge accumulates on the conductor surface, causing the electric field inside the conductor to be zero

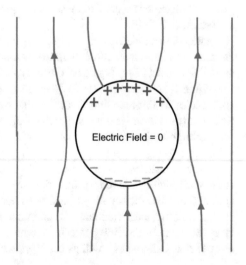

accumulate on the tissue surface. But if electrodes are present, the cathode draws those charges off the surface and the anode replenishes the charges somewhere else on the surface. The result is that charge constantly moves from the anode to the cathode. Therefore, you can have an electric field in your body if you pass current through electrodes, but not if you simply place your body in an external electric field.

Electroreception

Animals that live in the ocean do not shield themselves from electric fields as terrestrial animals do. If an electric field were present inside a saltwater fish, it would cause ions to move, but they would not remain on the fish's surface. Instead, they would flow into the surrounding seawater. We can say this in a different way: terrestrial animals live in air, which is an insulator, while fish in the ocean live in salt water, a conductor [3]. Not only does this mean that electric fields can penetrate into an aquatic animal, but also it implies that electric fields extend into the salt water and provide useful information about the surroundings. Some fish—including sharks, skates, and rays—have developed the ability to sense these weak electric fields: Electroreception.

One of the first studies of electroreception was performed by Adrianus Kalmijn, a biophysicist at the Woods Hole Oceanographic Institute. Kalmijn's experiments analyzed sharks (the kind that swim, not the kind that invest in entrepreneurial startup companies) [4]. He observed dogfish sharks while sitting in an inflatable rubber raft in the ten-foot-deep water of the Atlantic Ocean near Martha's Vineyard. He attracted the sharks using liquified herring placed on the ocean floor. On either side of the herring was a pair of electrodes that could pass current. The dogfish were initially lured by the smell of the herring, and "began frantically searching over the sand, apparently trying to locate the odor source" [4]. But when current was turned on, the dogfish stopped searching for the smell and viciously attacked the electrodes. With experiments like these, Kalmijn was able to show that sharks can sense a weak electric field.

Sharks detect electric fields using sensory organs in their snout known as ampullae of Lorenzini, which allow them to sense fields as small as 1 μV/m (Fig. 3.2). The ampullae consist of highly conducting jelly-filled tubes about 0.3 meters (roughly a foot) long. The shark detects the voltage across the length of the tube (there is an extra factor of three arising from the distortion of the field by the shark [5]), and then focuses this entire voltage, 1 microvolt, across a neuron's cell membrane.

A change in the membrane voltage of 1 microvolt is tiny. Recall that the typical membrane voltage is almost 100 millivolts, so this change is about ten parts per million. How can such a paltry voltage change be detected? Physicist William Pickard, from Washington University in Saint Louis, has analyzed this question [5]. He assumed that the membrane voltage does not make a neuron fire (it is far too weak for that), but instead modulates its spontaneous firing rate. Under idealized conditions, changes in the firing rate for a network of neurons have been observed for electric fields almost as small as 0.1 V/m, which is about a hundred times less than

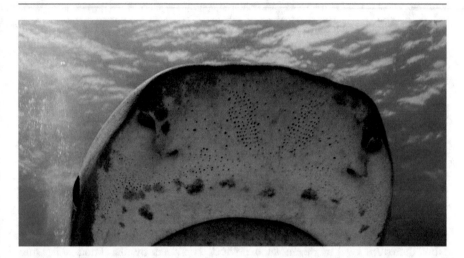

Fig. 3.2 Ampullae of Lorenzini pores on the snout of tiger shark. (Albert kok, CC BY-SA 3.0, https://creativecommons.org/licenses/by-sa/3.0, via Wikimedia Commons)

the threshold to stimulate a neuron [6]. Pickard's model requires that the neuron's firing rate be exceptionally sensitive to the membrane voltage, which magnifies a small change in voltage into a large change in rate.

Many ampullae of Lorenzini contribute to each neuron. This integration results in averaging the signal, thereby suppressing any background noise. For instance, in Chap. 2 we discussed how random thermal fluctuations set a limit to how small of a signal can be detected. Yale physicist Robert Adair has estimated that the size of thermal fluctuations of a neuron's membrane voltage are about 1 microvolt [7], comparable to the voltage produced by the weakest electric field a shark is able to detect. This estimate might be too low. In addition to thermal fluctuations, ion channels in the membrane randomly open and close causing the membrane voltage to fluctuate. Also, ions pass through the membrane one at a time (the current is discrete, not continuous), resulting in another source of fluctuations, called shot noise. When a neuron integrates the signal from hundreds of ampullae, the averaging tends to suppress all these fluctuations, allowing the system to pick the signal out of the noise. Shark electroreception appears to operate near the limit imposed by fluctuations. Evolution has honed a sensory mechanism that is about as sensitive as it can be without detecting the constant roar of random noise.

Electroreception may provide a way to navigate using the earth's magnetic field. We saw in Chap. 2 that when an organism moves in a magnetic field, it produces a weak electric field. If the shark is cruising at a leisurely 1 meter per second in a direction perpendicular to the earth's 50 μT magnetic field, it induces a 50 μV/m electric field, which is 50 times greater than the smallest field it can detect with its electroreceptor. Kalmijn has performed several experiments that suggest sharks take advantage of motion through the earth's magnetic field for navigation.

In Chap. 2 we learned that some animals are able to detect small, static magnetic fields, but these fields have little or no physiological impact on most animals. In this chapter, we found that a few animals can sense small, static electric fields. Next, we need to determine if electric fields affect the physiology of animals, including humans.

Bone Healing

Steady electric fields have been used to heal broken bones (Fig. 3.3) [8]. This work was motivated by bone's piezoelectric property: a mechanical stress produces an electric field, and conversely an electric field produces a mechanical stress. When orthopedic surgeons discovered bone's piezoelectricity, they wondered if electric fields could cause effects in addition to mechanical ones. In particular, they speculated about electric fields causing bone healing. The original experiments were promising, and doctors began to utilize electrical bone growth stimulators clinically. Now there are over a dozen such devices approved by the Food and Drug Administration for mending long bones (particularly those fractures that failed to heal on their own, called nonunions), and for fusing vertebrae in the spine.

Initial experiments on bone healing in the 1950s used a direct current (DC) electric field, meaning the field was static. This method requires that electrodes be implanted surgically near the fracture. Bone growth was observed at the cathode (negative electrode) and bone absorption at the anode (positive electrode). To avoid absorption, the anode was generally placed in the surrounding tissue, away from the bone itself. Most publications describing the use of DC electric fields for bone healing report the applied current as about 10 microamps. Converting the current to electric field strength is not simple, but the field appears to be on the order of 0.1–1 V/m [9]; not strong enough to excite neurons, but nearly a million times stronger than the weakest field a shark can detect.

When a metal electrode is in contact with the salt water making up much of our tissue, an electric current results in chemical reactions at the electrode surface (these reactions are necessary in order to transform current carried by electrons in the electrode wires into current carried by ions in the tissue). For alternating current, these often reversible reactions first progress in one direction and then the other, with little net effect. But with a direct current, the products of these reactions accumulate. Researchers believe that these products, rather than the electric field itself, are what stimulate bone growth. The cathode releases hydroxyl radicals, hydrogen

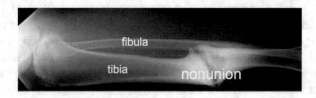

Fig. 3.3 A nonunion of a broken tibia (shinbone). (Lindsaydavidson, CC BY 3.0, https://creative-commons.org/licenses/by/3.0, via Wikimedia Commons)

peroxide, and other chemicals, which are thought to upregulate growth factors, thereby increasing bone deposition [8].

A DC electric field has several disadvantages. If too many reaction products amass, they may be toxic. Too much or too little of these chemicals is bad; the amount must just right. Also, implanting or removing the electrodes requires surgery, which is invasive, is usually painful, might cause infection, and could trigger inflammation.

To avoid surgery and other drawbacks, researchers developed two new methods for producing the electric field, based on capacitive and inductive coupling [8]. This chapter is focusing mainly on static electric fields, but to avoid spreading the story of bone healing over several chapters, I describe these electromagnetic stimulation methods now.

Both of these techniques use an alternating electric field or brief pulses. These approaches do not require chemical reactions at the surface of an electrode; for capacitive coupling the electrodes are on the skin well away from the bone, and for inductive coupling there are no electrodes at all. Therefore, these methods have a mechanism of interaction with bone that is different from DC stimulation. The healing is caused by the electric field itself, and not by a chemical reaction at the electrode surface. The details are uncertain, but any therapeutic response appears to derive from the electric field causing calcium ions to either enter the cell through channels in the membrane or be released from intracellular stores, initiating a complex biochemical respose [8]. The electric field strength employed by these instruments varies widely, but is generally stronger than for DC fields.

Does an electric field (alternating or steady) improve healing in patients with a broken bone? As in Chap. 2, researchers have performed many clinical trials, and the results are mixed. Several meta-analyses have attempted to sort through all the data [10–13]. For example, Brent Mollon and his coworkers examined 11 randomized clinical trials, nine of which were blinded to a placebo control. Their analysis was limited to electromagnetic stimulators. They found that "while our pooled analysis does not show a significant impact of electromagnetic stimulation on delayed unions or ununited long bone fractures, methodological limitations and high between-study heterogeneity leave the impact of electromagnetic stimulation on fracture-healing uncertain" [10]. Other meta-analyses, like that of Griffin and Bayat [11] and Aleem et al. [12], compared the clinical effectiveness of all three types of devices—DC, capacitive, and inductive—and were cautiously supportive of using electric fields for bone healing. The meta-analysis by Kooistra et al. [13] concluded that "although preclinical and observational evidence seems to provide a sensible rationale for using [electrical stimulation] in the treatment of long bone nonunions, the current paucity of and heterogeneity in sound clinical evidence prevent orthopedic surgeons from justifiably implementing it."

What do I make of these results? After decades of research and clinical use, we still are not sure if electric fields help bones heal. The evidence is so puzzling that even meta-analysis cannot make sense of it. Maybe electric fields do help, but maybe they do not. Nevertheless, when half a century of study is not sufficient to determine if a method works, I get suspicious. The technique is plausible, but definitely not proven.

Transcranial Direct Current Stimulation

Now that we have learned about the physics of electricity, discussed how sharks detect weak electric fields, and assessed how electric fields may or may not help bones heal, we are ready to analyze LIFTiD, the brain stimulation device featured on *Shark Tank*.

Experiments performed in the late 1800s by American physician Roberts Bartholow, Italian neuropsychiatrist Ezio Sciamanna, and Italo-Argentine neurosurgeon Alberto Alberti established that electrical stimulation can excite neurons in the human brain [14]. These researchers studied people who suffered from an injury or disease, such as cancer of the scalp and skull, that exposed their brains. When they passed current through electrodes placed on the brain surface, different parts of the body moved. If they stimulated the right side of the brain, a muscle on the person's left side contracted, and vice versa. During these experiments the patients were awake and claimed to feel no pain. Occasionally, however, stimulation led to a seizure.

Subsequent experiments on gorillas and other animals were performed by English physiologists such as Charles Sherrington. They mapped what regions of the brain controlled which body movements. Sherrington's student Wilder Penfield advanced this research further. Penfield was an American-Canadian neurosurgeon who treated epilepsy patients by removing parts of their skull and, while they were awake, probing the brain with electrical stimulation, searching for the origin of the seizure [15]. Once he found the source, he could surgically remove that part of the brain, curing the epilepsy. In the process, Penfield learned much about which regions of the brain control what functions of the body, including mental processes such as hallucinations and illusions.

This operation was useful if the brain was already exposed, or if curing a debilitating disease such as epilepsy could justify removing the scalp and skull. But could electrical stimulation affect brain function noninvasively: without any surgery? This is the goal of transcranial direct current stimulation, the approach adopted by the LIFTiD device (Fig. 3.4).

Transcranial direct current stimulation (often referred to as tDCS) has been proposed as a treatment for recovery from stroke, for mental illness, and for cognitive enhancement (making you smarter) [16]. The goal of many early studies of tDCS was to treat psychiatric disorders such as depression. Investigators claimed that anodal stimulation (using an electrode with a positive voltage) reduced depression, while cathodal stimulation (using a negative voltage) diminished mania. The response to 10 minutes of transcranial direct current stimulation generally lasted for an hour after the current was turned off. The voltage across a cell membrane typically decays in a few thousandths of a second, so tDCS must do more than merely

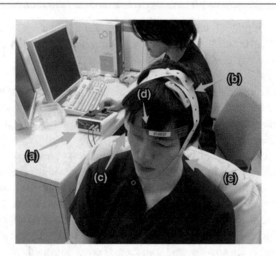

Fig. 3.4 Transcranial direct current stimulation, performed at the National Center of Neurology and Psychiatry Hospital in Tokyo, Japan. A subject (front) sits on a sofa relaxed, and a researcher (behind) controls the tDCS device (a). In this picture, an anode (b) and cathode (c) are attached to the scalp. The head strap (d) is for convenience and reproducibility, and a rubber band (e) is for reducing resistance. (Yokoi and Sumiyoshi. 2015, CC BY 4.0, https://creativecommons.org/licenses/by/4.0, via Wikimedia Commons)

vary the membrane voltage. Instead, longer-lasting changes in the brain structure (plasticity) are required.

A limitation of transcranial direct current stimulation is it cannot stimulate only a small area. Current passing through the skull tends to spread out, so it excites not only the neurons directly under the electrode but also those several millimeters on each side of it. Using a small electrode on the scalp does not result in stimulation of a limited region in the brain. In contrast, experiments performed on the exposed brain by Bartholow, Sciamanna, and Alberti, with the scalp removed, could use tiny electrodes to activate small regions of the brain, allowing for precise maps of brain activity.

Is transcranial direct current stimulation safe? Michael Nitsche and an international team of neurologists have examined tDCS and identified several potential safety hazards [16].

1. Reactions at the electrode surface can create toxins (like they did during bone healing using DC current), but these chemicals are released onto the skin, not into the brain. In addition, usually a saline-soaked sponge is placed between the electrode and the scalp, so the body is not in contact with the toxins at all.
2. In theory, tDCS could excite neurons deep in the brain that control breathing or the heartbeat, but in practice this does not happen.
3. Any metallic object—for example, a metal clip implanted during a previous brain surgery—could distort the electric field produced during tDCS, and patients with such implants should not undergo the procedure.

4. Any electric stimulation could potentially trigger a seizure, so patients with epilepsy should avoid brain stimulation. Fortunately, seizures have not been observed in healthy people.

Despite these possible risks, thousands of patients have undergone tDCS with only a few minor side effects, such as a headache. The technique is considered safe. The risks are low enough, and the technology is simple enough, that do-it-yourselfers have begun building home-made tDCS devices [17].

Typically, transcranial direct current stimulation uses a current of about 0.001 amp (1 milliamp). This current comes out of the anode and goes into the cathode. When one milliamp of current is applied through tDCS electrodes, the voltage difference between the cathode and anode is about 9 volts. Therefore, undergoing tDCS is similar to pressing the leads of a nine-volt battery against your head.

How strong of an electric field does transcranial direct current stimulation produce in the brain? To answer this question, Pedro Miranda, a physicist at the University of Lisbon, and his colleagues created a mathematical model [18]. A realistically-shaped head was divided into five regions: the brain (with its two types of tissue: white matter and grey matter), the cerebrospinal fluid surrounding the brain, the skull, and the scalp. They adopted the finite element method to solve the equation governing the electric field. This method is a computational technique that divides the head into two million volume elements, each having the shape of a tetrahedron (a triangular pyramid). The solution of the equation in each element is simple, but all the elements need to be pieced together, which is a gigantic

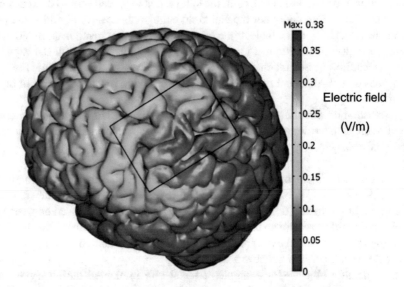

Fig. 3.5 The strength of the electric field on the surface of the brain during tDCS. The box indicates the position of the anode on the scalp. The current is 1 milliamp. (From Miranda et al. (2013) *NeuroImage* 70:48–58, with permission from Elsevier [18])

calculation. The computations took a couple hours, but when complete they provided the electric field distribution in exquisite detail (Fig. 3.5).

When one milliamp of current is applied through a scalp electrode, the largest electric field in the brain was calculated to be less than 0.5 V/m. The threshold for excitation of a neuron in the brain is about 10 V/m. This means transcranial direct current stimulation produces a voltage across the cell membrane with a strength about one twentieth of that needed to excite a neuron. Earlier, we found that electric fields in the brain as weak as about 0.1 V/m could, under ideal conditions, modify the rate of spontaneous neural activity [6]. Something similar to this must be occurring if tDCS is truly affecting the brain.

In 2009, I wrote an editorial in the journal *Clinical Neurophysiology* about another of Miranda's articles. The editorial identified the smallness of the electric field as an unresolved problem when trying to interpret tDCS experiments. I concluded that "Miranda et al.'s impressive calculations provide valuable insight into the electric field distribution induced during transcranial direct current stimulation. This same kind of meticulous modeling is required to determine how the neuron responds to the electric field" [19]. This conclusion remains true today.

Meta-analyses of tDCS

Although the detailed mechanism of how the brain responds to transcranial direct current stimulation is unknown, Miranda et al.'s calculation suggests the technique might barely be able to influence the rate of neuron firing. Let us adopt a pragmatic attitude; who cares how tDCS happens. The crucial question is: does it work? To answer this question, we look once more to meta-analysis.

Three meta-analyses published from 2014 to 2016 concluded that the influence of tDCS on memory is either small, partial, or nonexistent. Lauren Mancuso, of the University of Pennsylvania, and her collaborators reviewed these meta-analyses [20]. They focused on studies that stimulated the left dorsolateral prefrontal cortex (a part of the brain located roughly under the left forehead) and found a "small but significant" response when stimulation is coupled with working memory training (exercises meant to improve memory skills).

Among the general public the word "significant" might seem to be a synonym for "important," but to a statistician the meaning of "significant" is different. All experiments contain variation, and one goal of statistics is to determine if a result is real or could be merely a random fluke. For instance, if you flip a coin four times (heads you win, tails you lose) and get all tails, this could be because the coin has tails on both sides (unfair!), or it could be bad luck. In statistics, you can calculate the probability of flipping four tails in a row; it is one chance out of 16. So, you cannot say for sure the coin is dishonest, but merely that there is only a 6% chance that your result arose randomly. In statistics, the probability of a result being obtained by chance is quantified by the "p-value." A general practice in medicine is to say that any result having a p-value less than 0.05 is so unlikely to have occurred by chance that it is "significant." If the p-value is greater than 0.05 then the result is "not significant," because the odds of it being a random fluctuation are too high. Nothing magic happens when you reach p = 0.05, but you have to draw the line somewhere. (Other parameters are also used for significance testing, making the field of statistics complicated.) So, when Mancuso et al. write that they found a small but significant response, that means they are fairly confident that the response is real. They are not saying the effect is big or important. A little, inconsequential result may nevertheless be statistically significant.

Mancuso and her team found that stimulation of the brain when not associated with memory training was not statistically significant after correcting for publication bias: the tendency for only positive results to be published. Often scientists do not bother publishing an article that does not show an effect (a negative result). Therefore, if you analyze the published literature, you experience a bias toward studies saying a response is real. Your conclusion might be different if you knew about all those experimental results filed in someone's drawer but never written up. Publication bias can be corrected for partially with statistical techniques, and when these were applied to Mancuso's data the conclusion about stimulation without training did not reach the level of significance. They concluded that "the primary [working memory] enhancement potential of tDCS probably lies in its use during training" [20]. Publication bias is real and might be particularly common when deciding whether or not to publish studies about the biological consequences of electric and magnetic fields.

In 2017 Jared Medina and Samuel Cason, of the University of Delaware, reexamined Mancuso et al.'s study, using something called "p-curve analysis" [21]. The idea is to analyze the p-values reported in experimental studies, looking for tell-tale trends. For instance, if researchers tend to continue their experiments until they reach a p-value of 0.05 or lower and then stop and publish the results, this would cause the literature to have a clustering of p-values just below 0.05; a practice sometimes called p-hacking. The analysis is more sophisticated than this, but in a nutshell the distribution of p-values provides an additional source of information about possible biases in experiments. Medina and Cason also discussed publication bias and a dangerous practice known as HARKing: Hypothesizing After the Results are Known. In general, scientists first form a hypothesis, or guess, and then conduct an experiment to test that guess. Another more biased approach is to first conduct an experiment, then examine which of your results are significant, and finally form hypotheses to justify those results. After analyzing the data as completely as possible—accounting for publication bias, HARKing, and other factors—they found "no evidence that the tDCS studies had evidential value" [21]. In other words, the studies they examined do not provide enough evidence to test the hypothesis. Or, to put it more bluntly, after exhaustively analyzing the literature, they still do not know if transcranial direct current stimulation helps working memory or not.

The story is not yet over. Hannah Filmer, a psychologist at the University of Queensland in Australia, and her coworkers have addressed various criticisms of transcranial direct current stimulation [22].

1. They responded to the suggestion that the electric field in the brain is too small to modify neuron behavior by claiming that animal studies provide a "proof of concept" that a weak electric field can influence neural function.
2. Although scientists do not have a complete understanding of the mechanism (how it works) for transcranial direct current stimulation, Filmer et al. assert that recent research is beginning to address this problem. Also, you do not need to know how something works in order for it to work.
3. Even though tDCS results are often difficult to reproduce, the authors believe that the studies are complex and you need to examine the methodological details to determine if differences in experimental design may explain the irreproducibility.
4. Filmer and her collaborators speculate that the variable size of responses to stimulation might result from the inherent variability between subjects (for example, differences in skull thickness).
5. They acknowledge that some experiments are designed poorly, but suggest ways to improve future studies.
6. Transcranial direct current stimulation results may have been oversold to the public, but the authors suggest that this says more about the media than about the science.

Filmer and her colleagues conclude that "the field of tDCS has its imperfections, and certainly may not live up to all of the hype. Despite this, there is significant promise for tDCS to produce meaningful advances in our understanding of the brain" [22].

What conclusion do I draw from all this debate? First, science is difficult. Large effects are relatively easy to discover, but small ones tend to be buried in noise. Second, a single study or analysis does not decide an issue. Over time a body of literature grows and clarifies what is going on. Finding the truth does not come from a eureka moment, but instead from a slow slog ultimately leading to a consensus among scientists. Given these two considerations, we simply will have to wait and see what happens.

There is a critical factor in examining transcranial direct current stimulation that we have still not addressed: the placebo. Could the placebo effect be partly responsible for how tDCS works?

The Placebo During tDCS

One vexing topic in transcranial direct current stimulation is the placebo. We discussed in Chap. 2 how crucial a placebo is for a clinical trial. If a medication is being tested, the placebo is a sugar pill with the same size, shape, color, and taste as that of the drug. A placebo for transcranial direct current stimulation is difficult to design, because the electric field it produces in the brain is about a hundred times smaller than the electric field in the scalp (Fig. 3.6). Most of the current follows the path of least resistance, which is through the skin and highly conducting scalp (that

Fig. 3.6 During transcranial direct current stimulation, the electric field (red curves) is large in the scalp, and relatively small in the brain

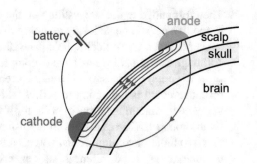

is, transcutaneous) rather than through the poorly conducting skull (transcranial). The relatively strong electric field in the scalp (often greater than 10 V/m) can excite pain receptors, peripheral nerves, and skeletal muscle. The titillation and annoyance resulting from stimulation of the scalp is presumably what caused the investors on *Shark Tank* to react with oohs and aahs when they tried out LIFTiD.

Luuk van Boekholdt, a graduate student at the Katholieke Universiteit Leuven in Belgium, and his coworkers analyzed this issue [23]. They noted that the usual sham (placebo) stimulation occurs by attaching the electrodes to the scalp and going through the same procedure as the patients receiving real stimulation, except the stimulus current is not turned on. That eliminates the weak electric field in the brain, which is presumably causing the clinical response, but also eliminates the stronger electric field in the scalp. In some cases, investigators deliver a placebo by turning the current on for a few seconds, and then turning it back off [16]. Most subjects claim to feel the current for only a short time when turned on or off. So, by turning current on only briefly, the perception in the scalp is similar with a placebo and with a long-lasting, say 20-minute, tDCS session.

Van Boekholdt postulates that two mechanisms exist that may be responsible for any tDCS clinical outcome: transcranial excitation of the brain and transcutaneous excitation of the scalp. Scalp stimulation could trigger behaviors, such as arousal and vigilance, that might be responsible for clinical outcomes such as treatment of depression or improvement in memory.

Van Boekholdt et al. suggest alternative sham experiments that investigators could take advantage of during tDCS to improve their research [23]. They propose applying an anesthetic cream to the scalp to deaden any perception of scalp stimulation (the cream would need to be applied to patients receiving either real or pretend stimulation). Another control could be putting the electrodes on some other part of the body, rather than the head. For instance, electrodes on the abdomen would provide a transcutaneous sensation, but would not produce an electric field in the head. This might be effective if patients are not aware that tDCS is supposed to work by stimulating the brain. All these placebos have their own drawbacks. Designing the ideal placebo is the Achilles heel of clinical trials using electrical stimulation.

Some practicing psychologists might not care whether the mechanism is transcranial or transcutaneous, as long as the treatment reduces depression in their patients. This is not an unreasonable attitude, but it takes the focus away from comprehending the mechanism of how transcranial direct current stimulation works. Such ignorance would make improving the technique difficult; if we do not know the mechanism, we do not know how to make the method better. Moreover, we would have to face all the tricky ethical questions we discussed in Chap. 2 when considering the placebo effect for magnets. In addition, that approach would contribute nothing to our understanding of how weak electric fields can affect the brain.

Ken and Allyson Davidov might have a difficult time marketing LIFTiD if it turns out that the mechanism has nothing to do with exciting the brain. As shark Mark Cuban said as he rejected their pitch for investment, Ken and Allyson "tried to sell science without using science" [1]. His conclusion might be more prescient than he realized. The science behind transcranial direct current stimulation is unclear. There are enough positive studies that we cannot reject it out of hand, but there are enough questions that we cannot accept it as an established therapy.

Chapters 2 and 3 examined the health consequences of static magnetic and electric fields. In many situations, however, these fields vary with time. In the next chapter, we examine if slowly varying fields, specifically 60 Hz fields associated with the power grid, are able to affect people.

References

1. https://www.dailymotion.com/video/x81b53e. Access date: January 12, 2020.
2. Feynman, R. P., Leighton, R. B., & Sands, M. (1964). *The Feynman lectures on physics* (Vol. 2). Addison-Wesley.
3. Denny, M. W. (1993). *Air and water: The biology and physics of life's media*. Princeton University Press.
4. Kalmijn, A. J. (1977). The electric and magnetic sense of sharks, skates, and rays. *Oceanus, 20*, 45–52.
5. Pickard, W. F. (1988). A model for the acute electrosensitivity of cartilaginous fishes. *IEEE Transactions on Biomedical Engineering, 35*, 243–249.
6. Francis, J. T., Gluckman, B. J., & Schiff, S. J. (2003). Sensitivity of neurons to weak electric fields. *Journal of Neuroscience, 23*, 7255–7261.
7. Adair, R. K. (1991). Constraints on biological effects of weak extremely-low-frequency electromagnetic fields. *Physical Review A, 43*, 1039–1048.
8. Khalifeh, J. M., Zohny, Z., MacEwan, M., Stephen, M., Johnston, W., Gamble, P., Zeng, Y., Yan, Y., & Ray, W. Z. (2018). Electrical stimulation and bone healing: A review of current technology and clinical applications. *IEEE Reviews in Biomedical Engineering, 11*, 217–232.
9. Tenforde, T. S. (1995). Spectrum and intensity of environmental electromagnetic fields from natural and man-made sources. In M. Blank (Ed.), *Electromagnetic fields: Biological interactions and mechanisms* (pp. 13–35). American Chemical Society.
10. Mollon, B., da Silva, V., Busse, J. W., Einhorn, T. A., & Bhandari, M. (2008). Electrical stimulation for long-bone fracture-healing: A meta-analysis of randomized controlled trials. *The Journal of Bone and Joint Surgery, 90*, 2322–2330.
11. Griffin, M., & Bayat, A. (2011). Electrical stimulation in bone healing: Critical analysis by evaluating levels of evidence. *ePlasty, 11*, 303–353.
12. Aleem, I. S., Aleem, I., Evaniew, N., Busse, J. W., Yaszemski, M., Agarwal, A., Einhorn, T., & Bhandari, M. (2016). Efficacy of electrical stimulators for bone healing: A meta-analysis of randomized sham-controlled trials. *Scientific Reports, 6*, 31724.

13. Kooistra, B. W., Jain, A., & Hanson, B. P. (2009). Electrical stimulation: Nonunions. *Indian Journal of Orthopaedics, 43*, 149–155.
14. Zago, S., Ferrucci, R., Fregni, F., & Priori, A. (2008). Bartholow, Sciamanna, Alberti: Pioneers in the electrical stimulation of the exposed human cerebral cortex. *The Neuroscientist, 14*, 521–528.
15. Ladino, D., Rizvi, S., & Téllez-Zenteno, J. F. (2018). The Montreal procedure: The legacy of the great Wilder Penfield. *Epilepsy & Behavior, 83*, 151–161.
16. Nitsche, M. A., Cohen, L. G., Wassermann, E. M., Priori, A., Lang, N., Antal, A., Paulus, W., Hummel, F., Boggio, P. S., Fregni, F., & Pascual-Leone, A. (2008). Transcranial direct current stimulation: State of the art 2008. *Brain Stimulation, 1*, 206–223.
17. https://www.scientificamerican.com/article/do-diy-brain-booster-devices-work. Access date: January 12, 2022.
18. Miranda, P. C., Mekonnen, A., Salvador, R., & Ruffini, G. (2013). The electric field in the cortex during transcranial current stimulation. *NeuroImage, 70*, 48–58.
19. Roth, B. J. (2009). What does the ratio of injected current to electrode area *not* tell us about tDCS? *Clinical Neurophysiology, 120*, 1037–1038.
20. Mancuso, L. E., Ilieva, I. P., Hamilton, R. H., & Farah, M. J. (2016). Does transcranial direct current stimulation improve healthy working memory? A meta-analytic review. *Journal of Cognitive Neuroscience, 28*, 1063–1089.
21. Medina, J., & Cason, S. (2017). No evidential value in samples of transcranial direct current stimulation (tDCS) studies of cognition and working memory in healthy populations. *Cortex, 94*, 131–141.
22. Filmer, H. L., Mattingley, J. B., & Dux, P. E. (2020). Modulating brain activity and behavior with tDCS: Rumours of its death have been greatly exaggerated. *Cortex, 123*, 141–151.
23. van Boekholdt, L., Kerstens, S., Khatoun, A., Asamoah, B., & McLaughlin, M. (2021). tDCS peripheral nerve stimulation: A neglected mode of action? *Molecular Psychiatry, 26*, 456–461.

Do Power Lines Cause Cancer?

4

In 1989, Paul Brodeur—an American science writer and longtime contributor to *The New Yorker* magazine—published the best-seller *Currents of Death* [1]. Brodeur made the case that electromagnetic fields associated with power lines, electric blankets, and video display terminals cause cancer. He wrote the book as a detective story, recounting the efforts of scientists such as Nancy Wertheimer and Robert Becker, who he cast as heroes in a battle to bring attention to alleged power line health risks. Furthermore, Brodeur insisted that those who disagreed with him were engaged in a conspiracy to hide the danger. "What can be done," he bemoaned, "about a problem whose very magnitude inspires either silence or denial among most officials of industry and government, and among so many members of the medical and scientific community?" [1] Electrical power is so pervasive in our society that if Brodeur's accusations were true the implications for public health would be enormous. Are the electromagnetic fields from power lines dangerous? Are scientists engaged in a cover-up of these dangers?

Frequency

Before tackling these frightening questions, we need to review some physics. The two types of electricity are direct current (DC) and alternating current (AC). We discussed direct current (also known as steady current) in the last chapter. Now we turn to alternating current. The electricity that powers our homes oscillates back and forth with a frequency of 60 times per second. The unit for frequency is the hertz (Hz), which is a cycle per second. It is named after the German physicist Heinrich Hertz, who was the first to experimentally detect electromagnetic waves in the 1880s (Fig. 4.1).

Sixty hertz (60 Hz) is a low frequency for electric and magnetic fields. For instance, the AM radio station 750 broadcasts its signal using electromagnetic fields oscillating at a frequency of seven hundred and fifty thousand hertz (750 kHz).

© The Author(s), under exclusive license to Springer Nature Switzerland AG 2022
B. J. Roth, *Are Electromagnetic Fields Making Me Ill?*,
https://doi.org/10.1007/978-3-030-98774-9_4

Fig. 4.1 Heinrich Hertz
(1857–1894). (Robert
Krewaldt, Public domain,
via Wikimedia Commons)

FM radio stations use frequencies of millions of hertz (MHz), and our cell phones utilize frequencies of billions of hertz (GHz). By comparison, a sluggish 60 Hz behaves almost like a static field. Sixty hertz electric and magnetic fields are often referred to as "extremely low frequency" electromagnetic radiation.

Not only our homes, but also our entire electrical power grid, use a frequency of 60 Hz. That means the electric current passing through those tall, gangly high-voltage power lines that link our cities to distant power plants oscillates 60 times per second (Fig. 4.2). Brodeur's claim is that 60-Hz power lines are dangerous and are capable of causing cancer. This allegation is related to the question about biological effects of steady electric and magnetic fields, which we discussed in Chaps. 2 and 3. However, 60 Hz fields have some unique features because they are changing with time.

Power Line Electric and Magnetic Fields

How big are the electric and magnetic fields associated with a power line? Thomas Tenforde of the Lawrence Berkeley Laboratory and William Kaune of the Pacific Northwest Laboratory estimated that the electric field under a high-voltage power line is as high as 15,000 V/m, and the magnetic field ranges up to 15 μT, or about a third of the earth's static magnetic field [2]. An electric field of 5000 V/m and magnetic field of 5 μT are more typical strengths associated with ordinary power lines.

The electric field produced by a power line may seem large, but it does not couple well to our body. Recall that surface charge shields our tissues and organs from a static electric field (Chap. 3). The same thing happens at extremely low

Fig. 4.2 High-voltage power lines. (Public Domain Dedication (CC0) Source: Unsplash)

frequencies, except the shielding is not perfect. The charge that builds up on the surface of the body oscillates 60 times a second, and as this charge sloshes back and forth it constitutes a current with an associated electric field. Nevertheless, at 60 Hz the shielding is still very effective, and the electric field inside us is small. On the other hand, magnetic fields are not shielded by the body; they pass through it with little or no attenuation. Because of this lack of shielding, when searching for biological hazards the power line magnetic field is the primary suspect.

When magnetic fields vary with time, a phenomenon we have not discussed yet enters our story: electromagnetic induction. In 1831, while working at the Royal Institution in London, English physicist Michael Faraday discovered that a changing magnetic field induces an electric field (Fig. 4.3). This landmark discovery underlies much of our modern electrical power technology.

As a result of induction, even if you were somehow completely shielded from the *electric* field produced by a power line, there would still be an electric field in your body induced by the changing *magnetic* field. Induction becomes increasingly important as the frequency gets higher. It is a minor effect for power lines, but will be critical for magnetic fields oscillating millions of times per second.

Fig. 4.3 Michael Faraday delivering a Christmas lecture at the Royal Institution on December 27, 1855. From a painting by Alexander Blaikley. (Public domain, via Wikimedia Commons)

Induced Power Line Electric Fields in the Body

How big are these induced electric fields? We can estimate them with a bit of physics. Assume the magnetic field has a strength of 5 μT and oscillates at 60 Hz. We also need to know the size of the induced current loop. The path of the current is complicated because our body does not have a simple shape. Let us take half a meter as a typical radius of the current loop induced by the changing magnetic field. In that case, the induced electric field has a strength of 0.0005 V/m. In this chapter, the electric fields in the body are so weak that we find it easier to express them in the units of *micro*volts per meter, so the induced electric field is 500 μV/m.

The induced electric field is small, but small is a relative term. What should we compare it to? The endogenous electric field—that field existing naturally inside your body—is generated by the beating of your heart. The endogenous field is what is measured when a doctor records your electrocardiogram (Fig. 4.4). This signal has an amplitude of about 0.001 volt (1 millivolt) when measured on the body surface using electrodes roughly half a meter apart, implying the electric field in your body is about 2000 μV/m. This is four times our estimate of the electric field induced by the power line magnetic field.

Does this estimation seem too simplistic? Rodney Hart and Om Gandhi, working at the University of Utah, have calculated the electric field associated with the electrocardiogram more accurately [3]. Using a realistically shaped model for the torso and appropriate conductivities for a variety of tissues, they found that the

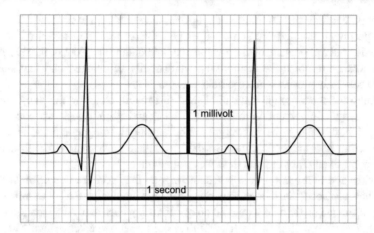

Fig. 4.4 The electrocardiogram

distribution of the endogenous electric field varies throughout the body: it is hetero-geneous. In the heart itself the field is large, about 3,000,000 μV/m. All through the torso it is about ten times greater than what we calculated previously. For instance, it is 40,000 μV/m in the intestines. Even in the brain the field strength is about 15,000 μV/m, seven times larger than our estimate. To be ultra conservative, we stick with our relatively low value of 2000 μV/m for the endogenous field, but keep in mind this is probably an underestimate, or lower bound.

Although the heart is the primary source of the endogenous electric field, it is not the only source. Electric fields are also produced by the brain, peripheral nerves, skeletal muscle, and the gut. Our calculation of the induced electric field is appro-priate near the edge of the body, where the current loop is largest. Deeper within the torso, the field is less. Finally, our assumed value of the magnetic field strength, 5 μT, is for directly underneath a power line. Magnetic fields in homes near a power line are usually about a tenth of this [2]. In other words, all our estimates made use of values that favor the induced electric field over the endogenous electric field, but still the endogenous electric field is bigger.

These calculations imply that electric fields induced in the body by power lines are smaller than the endogenous electric fields that always exist in our bodies. Granted, a power line field oscillates at a pure 60 Hz, whereas the endogenous field oscillates up and down in a range of frequencies. Nevertheless, we already see one problem with the idea that power lines cause cancer. The fields they create are over-whelmed by the larger electric fields that are always present in our tissues.

Can People Sense a 60 Hz Magnetic Field?

Robert Tucker and Otto Schmitt, working in the Electrical Engineering department at the University of Minnesota in the 1970s, decided to test if people could sense a 1000 μT, 60 Hz magnetic field (this field is 200 times greater than a typical field from a power line, and 20 times stronger than the earth's static field) [4]. Their initial experiment revealed that some "super perceivers" were able to tell reliably if a magnetic field was present, with a p-value of 0.0000000000000000000000000000001. There was no doubt these results were statistically significant, and that subjects could accurately predict when the field was on.

Some researchers might have published this result, claiming that they had discovered magnetic perception in humans. Tucker and Schmitt were skeptical, wondering if these super perceivers might not be sensing the magnetic field, but instead be detecting non-magnetic clues. So they launched a five-year effort to rid their experiment of any hint that the field was turned on.

Tucker and Schmitt's article is fascinating, as they recount this effort in detail [4]. They replaced the human experimenter by a computer that determined randomly if the magnetic field would be on or off during each test. At a location too far away to interfere with their experiment they built a second electromagnet identical to the real one used to apply the magnetic field. With it, they could avoid any clues from the slight dimming of the building lights caused by turning on the large current through the coil, or any soft hum associated with movements of the coil, or the clatter linked to turning relays on and off. They built an "acoustically padded cabinet" to further isolate the subject from sound and vibration.

These efforts increased the p-value to 0.00000000000000000001, which still means detection is highly significant. Tucker and Schmitt redoubled their effort. They built a new plywood cabinet to house the experiment, weighing over 600 pounds. They constructed it using no iron nails or hinges that could interact with a magnetic field, but instead fastened the cabinet together with a few brass screws. The door was seated on a rubber gasket and sealed by pulling a slight vacuum. Their next rounds of tests, carried out in this cabinet, showed that a few people could still detect when the magnetic field was on, with a p-value up to a still-significant 0.0000000001.

Scientists must be persistent. Tucker and Schmitt proved their tenacity with their next modifications. They hung the cabinet from the roof by an aircraft bungee shock cord to isolate it from even the mildest vibrations. To prevent the cabinet from swinging, they placed four slightly inflated automotive inner tubes between the floor and the cabinet base. They reclamped the turns of the coil to prevent any intra-coil "buzzing." They draped the inside of the cabinet with sound absorbing material, and they shock-mounted the subject's chair. After this Herculean struggle, they performed their final experiment, and found that subjects could not detect a 60 Hz magnetic field. Their article concluded:

> A large number of individuals were tested in this isolation system with computer randomized sequences of 150 trials to determine whether they could detect when they were, and when they were not, in a moderate (7.5–15 gauss [750–1500 μT] rms [root mean square]) alternating magnetic field… In a total of over 30,000 trials on more than 200 persons, no significantly perceptive individuals were found, and the group performance was compatible, at the 0.5 probability level [a 50/50 chance expected for random guessing], with the hypothesis that no real perception occurred. [4]

The Tucker and Schmitt experiment teaches a vital lesson. Scientific experiments are difficult, and scientists must work doggedly to eliminate any subtle systematic errors. As experiments improve, effects apparent in preliminary studies may become smaller and smaller, until disappearing once all sources of error are gone. Removal of artifacts is especially important when trying to distinguish a weak response from no response at all. One reason the bioelectric literature is filled with inconsistent results may be that not all experimenters are as diligent as Robert Tucker and Otto Schmitt.

Nancy Wertheimer and Epidemiology

At about the same time that Tucker and Schmitt were testing if people could detect 60 Hz magnetic fields, Nancy Wertheimer was driving around Denver trying to find some explanation for why so many local children were suffering from leukemia. Wertheimer was an epidemiologist: a scientist who analyses the patterns of health and disease among large groups of people. She noticed that often the children's homes were located near power lines. Wertheimer collaborated with a physicist, Ed Leeper, who developed a "wiring code" to estimate the strength of the magnetic field the children were exposed to. In 1979, Wertheimer and Leeper published an article titled "Electrical Wiring Configurations and Childhood Cancer" in the *American Journal of Epidemiology* [5]. Their conclusion was that children exposed to high magnetic fields were three times more likely to suffer from leukemia than those exposed to low fields. Paul Brodeur highlighted Wertheimer's research in *Currents of Death* [1]. In 1988, David Savitz and his University of North Carolina colleagues used improved methods to reexamine the relationship between power line magnetic fields and cancer. They found results encouraging enough to suggest the power line/cancer link found by Wertheimer and Leeper is real.

The Body Electric

Another scientist featured in Brodeur's book was the orthopedic surgeon Robert Becker, author of *The Body Electric: Electromagnetism and the Foundation of Life* [6]. Becker's book tells about his research into the consequences of electric currents on organisms, which was carried out at the Veterans Administration Hospital in Syracuse, New York. He is best known for his studies of regeneration (such as the regrowth of a salamander's leg after it was cut off), and for his work using electric fields for bone healing (see Chap. 3). He remains to this day an influential figure among those who believe power lines are hazardous to our health.

In *The Body Electric*, Becker supported several ideas that are not accepted by mainstream science, including

1. Nerve axons can conduct signals in only one direction,
2. Static magnetic fields affect people's cognitive ability,

3. Extrasensory perception (ESP) and psychokinesis (for example, spoon bending) work via electromagnetic fields,
4. Acupuncture is a bioelectric phenomenon, and
5. Sixty hertz magnetic fields (which Becker calls "electropollution") are more dangerous than nuclear weapons and could lead to the extinction of the human species.

He also claimed that nerves are semiconductors: electricity in nerves is conducted not by ions moving through water, but instead by electrons moving as they do through doped silicon. Although Becker was an early proponent of bone healing using DC electric fields, he rejected the use of AC fields induced by electromagnetic induction for treating broken bones because he feared those AC fields might cause cancer. *The Body Electric* describes Becker's campaign to prove that 60 Hz magnetic fields are dangerous.

In addition to the work of Wertheimer and Becker, in *Currents of Death* Brodeur promoted the research of an Australian-born medical doctor at the University of California in Los Angeles, W. Ross Adey. Adey is known for his reports of "window" effects, in which a moderate electromagnetic field resulted in a biological response, but both stronger and weaker fields did not. Besides windows in electromagnetic field strength, he also recorded analogous windows, sometimes called resonances, for frequency. For example, Adey observed that a 56 V/m electric field influenced calcium release from a chick brain to a greater degree for frequencies of 6 and 16 Hz rather than at either a lower frequency (1 Hz) or a higher frequency (32 Hz) [7]. Brodeur also highlighted the studies of Carl Blackman, of the Environmental Protection Agency, who claimed that a combination of a static magnetic field (with a strength similar to the earth's field) and an extremely low frequency electric or magnetic field caused another type of resonance, resulting in an efflux of calcium ions from brain tissue [8]. Finally, Brodeur cites the work of Andrew Marino, a scientist and lawyer who coauthored with his mentor Becker the book *Electromagnetism and Life*. Marino observed an influence of extremely low frequency fields on the neurological, endocrine, and immune systems of mice, which affected their growth and survival [9]. (*Electromagnetism and Life* suggested that biological tissue may contain regions of superconductivity, where the electrical conductivity goes to infinity. Superconductivity has never been observed in a living organism.) All this research led to a public fear of power line electromagnetic pollution. Becker and Marino testified about their work at public hearings to determine the safety of a new high-voltage power line in upstate New York.

In the 1980s and 1990s, Americans became increasingly afraid of power line magnetic fields. Parents lobbied—and in some cases sued—to move power lines away from their children's school. In 1992 Eddie Murphy starred in *The Distinguished Gentleman*. The film's plot is based in part on 60 Hz magnetic fields giving small-town children cancer. *Microwave News*, originally a monthly newsletter and now a website (www.microwavenews.com), was founded by Louis Slesin to report on the health consequences of electromagnetic fields, with an emphasis on the danger of power lines. Congress passed the Energy Policy Act in 1992 to fund research aimed at clarifying the risks of power line fields.

Just as concern over power lines was reaching its peak in the early 1990s, the idea that 60-Hz magnetic fields are dangerous was challenged by physicists.

Thermal Noise

One of the first physicists to enter the fray was Yale physics professor Robert Adair, a member of the National Academy of Sciences who was known for his research on elementary particles called kaons and for his interest in the physics of baseball. In 1991, Adair published an article in the leading physics journal *Physical Review* investigating the possible mechanisms by which 60-Hz electric and magnetic fields could affect organisms [10]. He began by analyzing mathematically some of the mechanisms we have already discussed. Electric fields are highly shielded by conducting tissues. Adair estimates that the 60-Hz electric field inside our body will be 50 million times smaller than the electric field in the air. In other words, a 5000 V/m power line field in the air (that is, 5,000,000,000 µV/m) is reduced to a 100 µV/m field in our body.

Adair compares this electric field to that produced by thermal noise. All the molecules in our body, including the charged ions, bounce around randomly as a result of their Brownian motion. As these ions move about, they create a fluctuating electric field. The size of this thermal field is tricky to calculate because it depends on how you measure it. If measured over short distances, the thermal noise is relatively large. But averaged over longer distances, the electric field at one location cancels the electric field at another, and the net thermal field is relatively small. Likewise, if measured over short times the thermal noise is large, but when averaged over longer times it is small. Adair estimated that when averaged over a distance equal to the size of a cell (about 20 microns) and over a time equal to one oscillation of the 60-Hz field (17 milliseconds), the thermal electric field has a size of 20,000 µV/m [10]. When we compare the 100 µV/m in our body caused by the power line electric field to the 20,000 µV/m created by thermal noise, we conclude that the thermal field is 200 times larger than the field caused by the power line.

If the thermal noise is 20,000 µV/m, how does the shark detect electric fields as small as 1 µV/m, as discussed in Chap. 3? The key is that Adair emphasizes "cell-sized regions." Most biological processes occur in cells, and for these processes the 20,000 µV/m value for thermal noise is appropriate. However, the shark's electric field sensor, the ampullae of Lorenzini, consists of tubes about a third of a meter long. When the thermal electric field is averaged over such a long distance, one region cancels the other, so the thermal noise in the tube is dramatically suppressed compared to that experienced by a 20-micron cell. On the other hand, the thermal electric field experienced by a typical protein molecule, which is about one hundredth of a micron in size, is much greater than that experienced by a cell.

What about power line magnetic fields? Earlier in this chapter we estimated the electric field induced in the body by a 5 µT power line magnetic field to be approximately 500 µV/m, which is 40 times less than the 20,000 µV/m thermal field.

Adair makes an additional comparison. In Chap. 2, we found that an electric field is produced by motion in a static magnetic field. Driving a car 60 miles per hour in the earth's magnetic field (which you experience every time you cruise down an interstate highway) generates an electric field of about 1000 µV/m, which is twice that induced by a power line magnetic field. You have to be careful with this comparison, because the motion-induced electric field is steady (except when you accelerate or turn), not oscillating 60 times per second. Nevertheless, this observation reinforces the conclusion that electric fields induced by power lines are weak.

What electric field is needed to produce neural excitation? In the units of this chapter, you would need 10,000,000 µV/m to active a neuron. Although we are not worrying about stimulating nerves in this chapter, this value further underscores the contention that the electric field induced by a power line is minuscule.

We have estimated several electric field values in this chapter, and it is easy to get them mixed up. Below I summarize the most important values:

20,000 µV/m	Thermal noise
2000 µV/m	Endogenous
500 µV/m	Induced in the body by a 5 µT power line
100 µV/m	Produced in the body by a 5000 V/m power line

From these values we draw two conclusions:

1. The electric field induced by the power line magnetic field is five times larger than that produced by the power line electric field. This is another reason why so much of the discussion about power line hazards centers on the magnetic field.
2. The thermal noise electric field experienced by a cell and the endogenous field produced by the heart are both larger than the electric field induced by a power line. Power line fields are swamped by thermal noise.

Adair also examines the possibility of resonances in biology [10]. A resonance occurs when an external electric or magnetic field has a frequency that matches the intrinsic frequency characteristic of some biological process. The resonance would allow for averaging of the effect over many oscillations, which would reduce the system's susceptibility to noise. An averaging time on the order of 20 minutes might be necessary to increase the signal-to-noise ratio enough for a signal to be detected. Adair shows that for such a long averaging time, the two frequencies—external and intrinsic—must be accurately matched to benefit from resonance. In other words, if the external frequency is 60.000 Hz but the intrinsic frequency is 60.001 Hz, the two frequencies are so distinct that a resonance effect will not happen. Could we have been so unlucky as to choose our power line frequency precisely in resonance with one of our body's intrinsic frequencies? In order for the frequency of a biological oscillation to be so precise (in the language of filters, in order for the passband to be so narrow), the intrinsic oscillation must not be damped. This means the biological oscillator should oscillate millions of times before it decays. Random thermal motion makes such a lack of damping unlikely.

After considering thermal noise and resonance, Adair concluded that "there are very good reasons to believe that weak [extremely low frequency] fields can have no significant biological effect at the cell level—and no strong reason to believe otherwise" [10].

More Physicists

In 1992, another physicist, John David Jackson—author of the renowned textbook *Classical Electrodynamics* [11], which I struggled through in graduate school—examined historical data on the growth of electric power since 1900 and the corresponding data on cancer (Fig. 4.5) [12]. If, as Wertheimer and Leeper contended, power lines caused leukemia, then the incidence of leukemia should have increased during the twentieth century, as the United States embraced electricity. Jackson found that per capita residential consumption of electricity has grown dramatically throughout the century, but the rate of cancer, including specifically childhood leukemia, has been flat or even slightly decreasing in the same period. He concluded:

> The absence of any appreciable change in the national cancer incidence rates during a period in which residential use of electric power has increased dramatically shows that the associated stray 50- or 60-Hz electromagnetic fields pose no significant hazard to the average individual. [12]

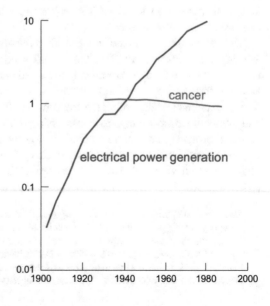

Fig. 4.5 United States per capita electric power generation (Megawatt-hour per year, red) and death rate for all cancers except respiratory (per thousand per year, blue). (Data from Jackson [12])

Radiation biologist John Moulder, of the Medical College of Wisconsin, began maintaining a website titled "Power Lines and Cancer FAQs [frequently asked questions]," which exhaustively summarized the evidence pro and con. Although this website is no longer available online, an archived pdf of it is [13]. In a 1996 article published by *IEEE Engineering in Medicine and Biology*, Moulder reviewed dozens of studies, and concluded that:

> Given the relative weakness of the epidemiology, combined with the extensive and unsupportive laboratory studies, and the biophysical implausibility of interactions at relevant field strengths, it is often difficult to see why there is still any scientific controversy over the issue of power-frequency fields and cancer. [14]

Finally, physicist Robert Park, the director of public information in the Washington DC office of the American Physical Society, lampooned the idea of power line fields causing cancer in his weekly newsletter *What's New*. I vividly remember when I was in graduate school how, every Friday, I would look forward to *What's New* appearing in my electronic mailbox. You can find many of Park's newsletters archived on the web [15].

The Stevens Report

The debate over the safety of power line magnetic fields led to a 1997 report by the National Academy of Sciences about "Possible Health Effects of Exposure to Residential Electric and Magnetic Fields" [16]. The chairman of the committee that drafted this report was neurobiologist Charles Stevens, of the Howard Hughes Medical Institute in La Jolla, California. The results of the committee's deliberations are usually called "The Stevens Report." Vice chairman was David Savitz, who had reproduced some of Wertheimer and Leeper's epidemiological results relating power lines to cancer.

The committee assessed over 500 published papers, including those by Wertheimer, Becker, Adey, Blackman, and Marino, as well as many subsequent articles that sought to reproduce and extend the work of those researchers. It assessed the exposure to power line fields, and the known mechanisms of interaction. It examined studies of signal transduction (how extracellular molecules regulate intracellular processes), gene expression (how our DNA instructs cells to make essential proteins), and calcium efflux. It reviewed reports on reproduction, behavior, and hormones. It considered experiments performed with cells, tissues, animals, and humans. It analyzed over a dozen epidemiological studies. The committee's report was published as a book with over 300 pages. It concluded that:

> Based on a comprehensive evaluation of published studies relating to the effects of power-frequency electric and magnetic fields on cells, tissues, and organisms (including humans), the conclusion of the committee is that the current body of evidence does not show that exposure to these fields presents a human-health hazard. Specifically, no conclusive and consistent evidence shows that exposures to residential electric and magnetic fields produce cancer, adverse neurobehavioral effects, or reproductive and developmental effects. [16]

And even though David Savitz was vice chair of the committee, it concluded that that "no evidence links contemporary measurements of magnetic-field levels to childhood leukemia" [16].

Later that year, the National Cancer Institute, part of the federal government's National Institutes of Health, published the results of a large, definitive epidemiological study of exposure to residential 60 Hz magnetic fields and cancer. Over 1200 children were divided into two groups: one with leukemia and another serving as the control. Magnetic fields in the children's homes were measured by technicians blinded to which of the two groups a particular child belonged. In an article published in the *New England Journal of Medicine*—one of the most respected medical journals in the United States—the authors reported that "our results provide little evidence that living in homes characterized by high measured time-weighted average magnetic-field levels or by the highest wire-code category increases the risk of [acute lymphoblastic leukemia] in children" [17].

The Stevens Report and the National Cancer Institute study essentially ended the debate over power lines and cancer. In his book *Voodoo Science*, Robert Park wrote that

> the supposed association between proximity to power lines and childhood leukemia, which had kept the controversy alive all these years, was spurious—just an artifact of the statistical analysis. As is so often the case with voodoo science, with every improved study the effect had gotten smaller. Now, after eighteen years, it was gone completely. [18]

References

1. Brodeur, P. (1989). *Currents of death: Power lines, computer terminals, and the attempt to cover up their threat to your health*. Simon & Schuster.
2. Tenforde, T. S., & Kaune, W. T. (1987). Interaction of extremely low frequency electric and magnetic fields with humans. *Health Physics, 53*, 585–606.
3. Hart, R. A., & Gandhi, O. P. (1998). Comparison of cardiac-induced endogenous fields and power frequency induced exogenous fields in an anatomical model of the human body. *Physics in Medicine and Biology, 43*, 3083–3099.
4. Tucker, R. D., & Schmitt, O. H. (1978). Tests for human perception of 60 Hz moderate strength magnetic fields. *IEEE Transactions on Biomedical Engineering, 25*, 509–518.
5. Wertheimer, N., & Leeper, E. (1979). Electrical wiring configurations and childhood cancer. *American Journal of Epidemiology, 109*, 272–284.
6. Becker, R. O., & Selden, G. (1985). *The body electric: Electromagnetism and the foundation of life*. William Morrow.
7. Adey, W. R. (1981). Tissue interactions with nonionizing electromagnetic fields. *Physiological Reviews, 61*, 435–514.
8. Blackman, C. F., Benane, S. G., Rabinowitz, J. R., House, D. E., & Joines, W. T. (1985). A role for the magnetic field in the radiation-induced efflux of calcium ions from brain tissue in vitro. *Bioelectromagnetics, 6*, 327–337.
9. Becker, R. O., & Marino, A. A. (1982). *Electromagnetism & life*. SUNY Press.
10. Adair, R. K. (1991). Constraints on biological effects of weak extremely-low-frequency electromagnetic fields. *Physical Review A, 43*, 1039–1048.
11. Jackson, J. D. (1999). *Classical electrodynamics*. Wiley.

12. Jackson, J. D. (1992). Are the stray 60-Hz electromagnetic fields associated with the distribution and use of electric power a significant cause of cancer? *Proceedings of the National Academy of Sciences, 89,* 3508–3510.
13. large.stanford.edu/publications/crime/references/moulder/moulder.pdf. Access date: January 12, 2022.
14. Moulder, J. E. (1996). Biological studies of power-frequency fields and carcinogenesis. *IEEE Engineering in Medicine and Biology Magazine, 15,* 31–49.
15. https://web.archive.org/web/20140124195058/http://bobpark.physics.umd.edu/archives.html. Access date: January 12, 2022.
16. National Research Council. (1997). *Possible health effects of exposure to residential electric and magnetic fields.* National Academy Press.
17. Linet, M. S., Hatch, E. E., Kleinerman, R. A., Robison, L. L., Kaune, W. T., Friedman, D. R., Severson, R. K., Haines, C. M., Hartsock, C. T., Niwa, S., Wacholder, S., & Tarone, R. E. (1997). Residential exposure to magnetic fields and acute lymphoblastic leukemia in children. *New England Journal of Medicine, 337,* 1–7.
18. Park, R. L. (2000). *Voodoo science: The road from foolishness to fraud.* Oxford University Press.

Will Electrical Stimulation Help Your Aching Back?

5

Clyde Norman Shealy was only 23 when he graduated from the Duke School of Medicine. During the 1960s, he worked at Western Reserve University (now part of Case Western Reserve) developing methods to control pain. He describes a case study in a preliminary clinical report published in *Anesthesia & Analgesia*:

> A 70-year-old man was admitted to Lutheran Hospital in early March because of severe diffuse pain in the right lower part of the chest and the upper part of the abdomen.... On March 24, 1967, a thoracic laminectomy [removal of the covering around the spinal cord]... was performed and a Vitallium [an alloy of cobalt, chromium, and molybdenum] electrode measuring 3 by 4 mm. was approximated to the dorsal columns at D3 [one of the vertebrae in your spine]... At 6 p.m. on March 24, [electrical] stimulation was begun with 10 to 50 pulses per second (400-msec. [perhaps they meant μsec.?] pulses, 0.8 to 1.2 volts, and 0.36 to 0.52 ma.). The patient detected a 'buzzing' sensation in his back which extended around and throughout his chest but not into his legs. *Both incisional and original pain were immediately abolished* [my italics]. [1]

Motivated by this success, Shealy continued his research into electrical stimulation. In order to avoid surgically implanting electrodes in the spine, he began to place them on the skin. This work led to what is now known as transcutaneous electrical nerve stimulation, or TENS for short. The Food and Drug Administration has approved TENS devices, and doctors often recommend them to control a variety of disorders, including back pain. You do not need a prescription, as TENS units are now sold over-the-counter. Typically, users can adjust the stimulus pulse duration, intensity, and frequency to meet their needs.

The transcutaneous electrical nerve stimulator is one example of a device that makes use of electricity to activate nerves or muscles. Other examples abound. One of the first and most successful is the cardiac pacemaker, which activates the heart.

© The Author(s), under exclusive license to Springer Nature Switzerland AG 2022
B. J. Roth, *Are Electromagnetic Fields Making Me Ill?*,
https://doi.org/10.1007/978-3-030-98774-9_5

Pacemakers and Defibrillators

During World War Two, Paul Zoll was a cardiologist assigned to a US army hospital, where he served on a surgical team [2]. As the surgeons removed shrapnel from soldiers' chests, Zoll noticed how easily the heart could be excited by the slightest touch. After the war, he returned to Beth Israel Hospital in Boston with the goal of inventing a cardiac pacemaker. In the fall of 1952, he tried out his prototype pacemaker to treat a patient with heart block (Fig. 5.1). Block occurs when the electrical connection between the upper and lower chambers of the heart is disrupted, so the cardiac electrical activity cannot pass from the atria (the upper, smaller chambers) to the ventricles (the lower, more powerful chambers), which causes the heart to not pump correctly and can lead to fatigue, dizziness, or fainting. A pacemaker stimulated the ventricles using an electric current, causing the cardiac muscle to contract at the appropriate rate. His device not only kept the desperately ill patient alive, but also allowed the patient to eat and listen to a radio broadcast of the World Series. The era of cardiac pacing had begun.

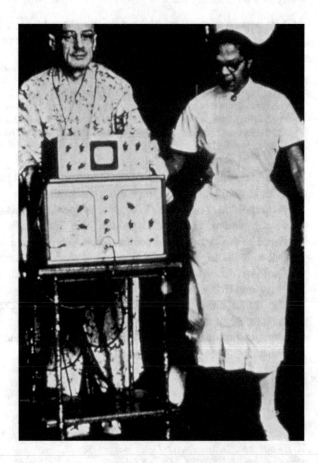

Fig. 5.1 The external pacemaker developed by Paul Zoll in 1952. (Reproduced with permission of Medtronic)

Zoll's first pacemaker had several problems. It was powered by being plugged into a wall outlet, so a patient could move no farther than an extension cord would allow. A power outage would be deadly. The stimulator was bulky and needed to be pushed around on a cart (Fig. 5.1). The electrodes were attached to the skin on the chest, so they stimulated not only the heart, but also skeletal muscles and nerves, which made pacing painful. The device was useful for treating acute (short term) heart block but was not practical for chronic (long term) illness.

An improved pacemaker was invented by Earl Bakken, an engineer who had started a medical equipment repair service in his Minneapolis garage [2]. To free the patient from the extension cord, Bakken powered his pacemaker with a 9-V battery. He based its design on an electrical circuit similar to one in an electronic metronome that generated a steady rhythm to maintain the correct tempo when playing the piano. He miniaturized the pacemaker using a transistor circuit (a new technology at that time), which was compact enough that it could be attached to a belt or hung from a strap around the neck. The electrodes were inserted through the skin and positioned near the heart, reducing the required stimulus current and avoiding muscle and nerve activation. This pacemaker not only could perform acute pacing, but also could treat chronic heart disease [3]. Bakken's garage business grew into Medtronic, a Fortune 500 company and the world's largest medical device manufacturer.

One of the early Medtronic pacemakers was used by a 72-year-old retired car salesman, Warren Mauston. For over 6 years, Mauston lived an active life with the pacemaker, despite having a wire passing through his skin connecting the device to his heart. He loved to regale reporters with his story, and a 1961 article about him in the *Saturday Evening Post* helped pacemakers become accepted by the public.

Bakken's design had one major drawback: the electrode passing through the skin opened up a path for infection. The next advance in pacemaker design was made by an electrical engineer who taught at the University of Buffalo, Wilson Greatbatch. In 1958, Greatbatch invented a pacemaker you could place entirely inside the body (Fig. 5.2) [2]. The device was about the size of a hockey puck and contained seven batteries and all the needed circuitry. In April 1960, heart surgeon William Chardack implanted one of Greatbatch's pacemakers into a 77-year-old patient suffering from heart block. The pacemaker kept the patient alive for another two and a half years [4].

Implantation of Greatbatch's original pacemaker required open heart surgery. Soon, however, doctors devised a way to snake the electrode lead into the heart through a vein, with the pacemaker itself placed under the skin in the upper chest. Today, such pacemakers are common, and having one implanted is routine (Fig. 5.3).

The big brother of the pacemaker is the defibrillator [2]. The heart normally beats in a regular fashion, but occasionally it can become chaotic, a disorder known as fibrillation. If the ventricles—the chambers of the heart primarily responsible for pumping blood—fibrillate, death follows in minutes. A defibrillator is similar to a pacemaker but is more powerful. An implanted defibrillator monitors the electrocardiogram, and when it detects fibrillation it delivers a shock that resets the heart to a normal beat. Automated external defibrillators are installed and maintained in

Fig. 5.2 Wilson Greatbatch's implantable pacemaker, invented in 1958. (From Chardack et al. (1960) *Surgery* 48:643–654 [4])

Fig. 5.3 A modern pacemaker. The device is implanted under the skin in the upper chest. A wire lead passing through a vein connects the pacemaker to the heart. (From: https://www.nhlbi.nih.gov/health-topics/pacemakers)

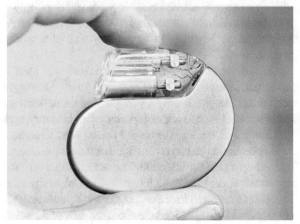

schools, churches, and on airplanes, available if needed, like a fire extinguisher. Companies such as Medtronic manufacture both pacemakers and defibrillators.

How big are the electric fields required for pacing and defibrillation? I have performed computer simulations of cardiac pacing and calculated that an electric field of about 100 V/m needs to be produced next to the stimulus electrode [5]. Defibrillation requires an electric field about ten times larger, 1000 V/m, throughout most of the heart. The energy depends on the square of the electric field, so defibrillation takes at least a hundred times as much energy as pacing, usually more. Both devices use pulses lasting about one millisecond. If we think of such a stimulus pulse as being made up of many different frequencies—a technique pioneered by French mathematician Joseph Fourier in the 1820s—then pacing and defibrillation use frequencies of around 1000 Hz (a kilohertz); a higher frequency than the 60 Hz of power lines, but a lower frequency than used by radio stations, television broadcasts, and cell phones.

Kirk Jeffrey's book *Machines in Our Hearts* [2] chronicles the history of the cardiac pacemaker and defibrillator, and describes the rise of the medical device industry. Pacemakers and defibrillators are two of the greatest inventions of modern medicine. They save lives.

How Nerves Work

Although the heart was the first organ to be stimulated electrically by a medical device, much recent research has concentrated on stimulating nerves. The biophysics of nerve electrophysiology has been understood since the groundbreaking research of English physiologists Alan Hodgkin and Andrew Huxley (Fig. 5.4) [6]. Hodgkin and Huxley worked at Cambridge University, but during the summers they performed experiments at the Plymouth Marine Laboratory. Their initial research on nerve electrical behavior was interrupted for 5 years because of World War II. Once they returned from war work, they renewed their studies on nerves, which ultimately resulted in their Nobel Prize for Physiology or Medicine in 1963.

A major reason for Hodgkin and Huxley's success was that they had access to a large nerve axon in the squid, a soft-bodied mollusk that lives in the Atlantic Ocean. An axon is the long, cylindrical part of the nerve fiber that transmits the electrical signal. A typical axon is about 20 microns in diameter or smaller. The squid giant

Fig. 5.4 Alan Hodgkin (1914–1998, left) and Andrew Huxley (1917–2012, right). (Public domain, via Wikimedia Commons)

axon, however, has a diameter of up to 1000 microns. This is big enough to insert electrodes into it, which allowed Hodgkin and Huxley to measure the voltage and current across the axon's membrane. You can watch Alan Hodgkin performing some of his pioneering experiments on the squid giant axon in a video on YouTube [7].

When a nerve is at rest, a voltage of about −70 millivolts exists across its cell membrane with the minus sign indicating that the inside of the nerve is negative. If a stimulus changes this voltage to make it less negative by about 20 millivolts (so, the voltage changes from −70 to about −50 millivolts), it opens channels in the membrane that let in positively charged sodium ions, making the inside of the nerve even less negative. The resulting positive feedback loop (a less negative voltage makes the sodium channels open more, allowing more sodium to enter the cell, making the voltage even less negative, etc.) causes the membrane voltage to surge explosively, until the inside of the axon becomes positively charged (the voltage rises to about +40 millivolts). This "action potential" lasts for a few milliseconds before the nerve returns back to rest.

An action potential is able to propagate down a long nerve axon. When one region of the axon fires an action potential, it excites a neighboring region, causing its sodium channels to open and fire an action potential. The process is analogous to a burning fuse; when one end is lit, it catches fire, heating the adjacent bit of fuse until it is hot enough to catch fire, and so on. This propagation of the action potential allows the electrical signals to be sent from one location in the body to another, such as from the brain to a leg muscle.

Hodgkin and his coworkers found a clever way to show that the cell membrane and its ion channels produce the action potential rather than any structure inside the cell. They removed the axoplasm (the intracellular material enclosed by the membrane) from a giant squid axon, using a small roller to squeeze it out, and replaced it by salt water (Fig. 5.5) [8]. Even with these changes, the action potential still propagated along the axon just fine. This experiment contradicts Robert Becker's claim that electricity inside nerves is not carried by ions moving in water, but instead is a mysterious form of semiconduction (see Chap. 4).

A key factor responsible for the action potential was the sodium that entered the cell through an ion channel [9]. These channels are large proteins in the nerve membrane with a central, water-filled pore. The structure of these ion channels has been determined in detail, down to the position of each atom. Scientists have also uncovered how and why channels are selective (only sodium ions pass through some, while only potassium ions pass through others). You can learn more about the structure and selectivity of ion channels by listening to Roderick MacKinnon's Nobel lecture on YouTube [10]. Genetic mutations that change an ion channel can lead to

Fig. 5.5 Internal perfusion of a squid nerve axon. The axoplasm (yellow) is squeezed out and replaced by salt water

diseases, called channelopathies. An unforgettable example is a myotonic goat: a goat that stiffens and falls over when startled, then later gets back up again and resumes normal life. This disorder is caused by a mutation in the chloride channels of the goat's muscles [11].

Scientists are able to study ion channel behavior in detail using patch clamping [9]. Erwin Neher and Bert Sakmann developed the patch clamp in the late 1970s. To patch clamp, a tiny glass electrode is pressed against a nerve to form a tight seal between the electrode tip and a micron-sized patch of membrane, allowing recordings of current through individual ion channels. In 1991, Neher and Sakmann received the Nobel Prize for this work.

Mathematical equations, including those of the influential Hodgkin and Huxley model [6], predict accurately the process of nerve conduction. In Chaps. 2, 3 and 4 of this book, researchers often did not know the mechanism underlying the biological response to weak electric and magnetic fields; they observed an effect but did not know what caused it. When studying the electrophysiology of nerve and muscle cells, the general mechanism underlying an action potential is now understood, although researchers may argue over the details.

Neural Prostheses

Neural stimulation can be employed to create all sorts of wonderful medical devices, including many prostheses. The goal of a prosthesis is to restore function to a part of the body that is impaired. For example, deafness usually arises as a result of injury to auditory hair cells in the inner ear that are responsible for converting sound into a neural signal sent to the brain. In many deaf people, the auditory nerve itself is intact and functioning, but the hair cells are damaged and not able to activate it. This nerve can be artificially stimulated with an auditory prosthesis known as a cochlear implant (Fig. 5.6) [12].

A cochlear implant has several parts: the microphone records sound waves, and a miniature computer splits the sound into its component frequencies (Fourier analysis again), then the computer directs small electrodes placed in the cochlea (the part of the inner ear that has a spiral shape like a snail shell) to stimulate the auditory nerve. Different locations in the cochlea correspond to different frequencies, and the computer determines which electrodes should be activated to create the sensation of sound at the appropriate pitch. The electric field required for neural stimulation is about the same as needed for pacing, roughly 100 V/m, or perhaps slightly less. The pulse duration is shorter, on the order of half a millisecond.

Hearing can be partially restored using a cochlear implant. The quality of sound is not perfect; patients with these implants might not be able to appreciate the subtleties of Beethoven's music but they might be capable of carrying on a conversation. For many, that ability is life-changing.

Other research into developing neural prostheses has been performed at Case Western Reserve University in Cleveland. Much of this research was led by Case

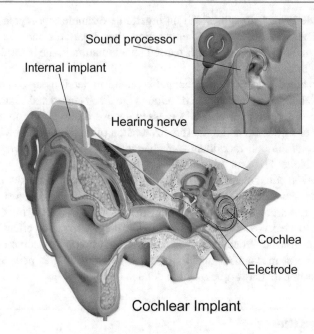

Fig. 5.6 A cochlear implant. The microphone (brown disk) records the sound outside of the ear, and the sound processor (a mini computer) transforms the sound into instructions for electrical stimulation of the auditory (or hearing) nerve. The internal implant has a wire that connects to an electrode in the cochlea. (From: Blausen.com staff (2014). "Medical gallery of Blausen Medical 2014". WikiJournal of Medicine 1 (2). DOI:https://doi.org/10.15347/wjm/2014.010. ISSN 2002-4436., CC BY 3.0, https://creativecommons.org/licenses/by/3.0, via Wikimedia Commons)

engineer J. Thomas Mortimer [12]. When growing up in Texas, one of Mortimer's close high school friends was paralyzed in an auto accident. This disaster motivated Mortimer to develop functional neural stimulators. After training under C. Norman Shealy, Mortimer established and directed Case's Applied Neural Control Laboratory.

P. Hunter Peckham, who was Mortimer's first graduate student, developed a prosthetic device to restore function to a patient's paralyzed hand. Another Case researcher, Ronald Triolo, invented a stimulator that allowed a wheelchair-bound patient to stand and walk. Quadriplegics often have difficulty controlling their urination and bowel movements. Mortimer and his colleague Graham Creasey developed a prosthesis to control the bladder and bowel muscles. Victor Chase has told the story of how these neural prostheses were developed in his book *Shattered Nerves*. He concludes that "neural prostheses have the potential to aid hundreds of thousands, or perhaps millions, of individuals with neurological disorders" [12].

Fig. 5.7 An x-ray of implanted deep brain stimulation electrodes. (Jmarchn, CC BY-SA 3.0, https://creativecommons. org/licenses/by-sa/3.0, via Wikimedia Commons)

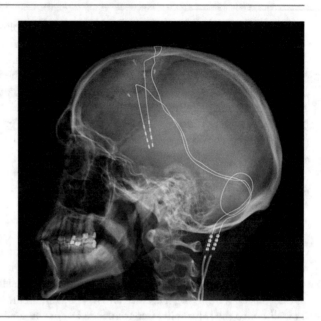

Brain Stimulation

Electrical stimulation of the brain is not as well understood as stimulation of peripheral nerves, such as the ones Hodgkin and Huxley studied. Nevertheless, techniques have been developed to excite neurons in the brain (Fig. 5.7). For example, deep brain stimulation is a treatment to reduce tremor (an involuntary twitching of muscles) [12]. The method is illustrated by the amazing case of the musician Roger Frisch. After performing in the Minnesota Orchestra for decades, Frisch developed a tremor that made his hands shake so that he could not play his violin. When doctors at the Mayo Clinic electrically excited neurons using electrodes implanted in Frisch's brain, the tremor ceased. Because the electrodes must be placed at precisely the correct location, Frisch was required to remain awake during the implant surgery. He played his violin while the operation was going on, so that the doctors could observe his ability to control his hands as they moved the electrode around. You can find a fascinating video of Frisch's operation on YouTube [13]. Since then, deep brain stimulation has become a common treatment for patients with Parkinson's disease.

The electric field strength required for deep brain stimulation is similar to that for pacing the heart: 100 V/m. Brain stimulation, however, uses briefer pulses; pacemaker pulses last about one millisecond, whereas deep brain stimulation pulses are shorter, about a tenth of a millisecond. In both cases, the electric field falls off with distance, so only tissue next to the electrode is excited. That is why positioning the electrode in the brain correctly is so crucial.

Another, more violent, type of brain stimulation is electroconvulsive therapy, which is used to treat patients with severe depression. A large pulse of current is passed through electrodes attached to the scalp (a current of about 1 amp, and a pulse duration of about one millisecond). The electric field produced in the brain is comparable to that used during deep brain stimulation, about 100 V/m. However, a much larger region of the brain must experience this field. Electroconvulsive therapy generally triggers a seizure and is performed under anesthesia. A meta-analysis, performed by Daniel Pagnin and his collaborators at the University of Pisa, analyzed several randomized, controlled clinical trials and found that electroconvulsive therapy is significantly better than a placebo for treating depression and is significantly better than antidepressant drugs [14]. Despite the procedure's horrifying portrayal in *One Flew Over the Cuckoo's Nest*, it has proven to be effective.

Electric Fish

We have been examining medical devices that produce strong electric fields, but some animals can create strong fields too. The electric eel is a freshwater fish found in the Amazon that stuns its prey with a powerful electric shock (Fig. 5.8). The eel acts like a taser, stimulating nerves in a fish it targets, which causes the fish's muscles go into tetanus, incapacitating it [15]. The eel can generate a 500 volt shock and is a meter long, implying it creates an electric field on the order of 500 V/m, which is well above the threshold strength needed to excite a nerve and is similar to that needed to initiate a heart arrhythmia. By curling around its prey, it makes an even larger electric field. The Tennessee Aquarium's electric eel, Miguel Wattson, has his own Twitter account (@EelectricMiguel) that sends out a tweet every time he delivers a shock. An electric ray, known as a torpedo, also shocks its prey. Scribonius Largus, the physician to the Roman emperor Claudius, used the torpedo to treat headaches and gout.

Fig. 5.8 An Electric Eel. (From: Steven G. Johnson, CC BY-SA 3.0, http://creativecommons.org/licenses/by-sa/3.0, via Wikimedia Commons)

Fig. 5.9 Transcranial magnetic stimulation. A current pulse in the circular coil (yellow) produces a changing magnetic field (red) that induces an electric field in the brain (green). (Eric Wassermann, M.D., Public domain, via Wikimedia Commons)

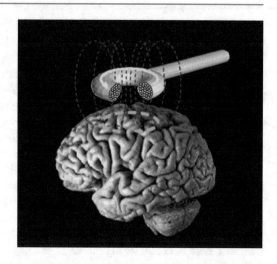

Transcranial Magnetic Stimulation

Chapter 4 introduced Michael Faraday's concept of electromagnetic induction: a changing magnetic field induces an electric field. Researchers have developed a technique based on induction to stimulate neurons in the brain: Transcranial magnetic stimulation (Fig. 5.9). This method has several advantages. When the brain is stimulated with electrodes attached to the head, the electric field in the scalp is much greater than the field in the brain (recall that the same thing happened during transcranial direct current stimulation discussed in Chap. 3). However, magnetic fields are not attenuated by the body. Therefore, transcranial magnetic stimulation activates the brain without a stronger, potentially painful, electric field being induced in the scalp. Deep brain stimulation requires brain surgery to implant the stimulating electrode, but transcranial magnetic stimulation is noninvasive. The doctor passes a pulse of current through a coil held near the head. This current produces a magnetic field that induces an electric field in the brain. There is no surgery, nothing is implanted, and the coil and head are not even in physical contact.

When I worked at the National Institutes of Health in the early 1990s, I calculated the electric field induced inside the brain during transcranial magnetic stimulation [16]. The induced electric field is approximately 100 V/m, a value familiar to you by now. Doctors would love to be able to restrict that field to a limited area. As with deep brain stimulation, they often want to activate only a small region of the brain. A disadvantage of transcranial magnetic stimulation is that the induced electric field is not localized. A well-designed coil can reduce the extent of the activated region to about a centimeter in size [16]. The strength of the magnetic field created in the head is approximately 1 T, or nearly as large as exists during magnetic resonance imaging.

Little of the energy generated in the coil is transferred to the body. Huge currents, thousands of amps, pass through the coil in order to induce weak currents in the brain. If such a giant current flowed continuously, the coil would melt. Fortunately, the current pulse lasts only about a tenth of a millisecond, so the copper in the coil winding does not have enough time to heat up. Nevertheless, heating is a problem during rapid rate transcranial magnetic stimulation, when 30 pulses are applied each second.

One reason so little energy is absorbed by the body is that the electrical conductivity of copper is more than a million times greater than the conductivity of the brain. If magnetic stimulation were applied when a good conductor was nearby, large currents would be induced in the conductor. For instance, researchers have performed transcranial magnetic stimulation while simultaneously recording the electroencephalogram using electrodes on the scalp. These electrodes are usually made of silver, an exceptional conductor. During the treatment, such large currents are induced in the silver that the electrode gets hot, and may burn the subject's scalp [17]. Moreover, the induced currents in a conductor may interact with the large magnetic field to create strong forces. If you place a coin in the center of a coil and then stimulate, the coin shoots upward toward the ceiling. Transcranial magnetic stimulation is safe only because so little of the enormous energy in the coil is transferred into the brain.

Transcranial magnetic stimulation is an excellent research tool. It can map out which part of the brain is responsible for a particular function. Researchers at the National Institutes of Health have used magnetic stimulation to determine how the brain changes as we learn, a process called plasticity. Because the technique is painless and noninvasive, they are able to apply it repeatedly and discover, for instance, how the brain adapts as a blind person learns to read braille. It is also used to diagnose diseases such as multiple sclerosis, which are caused by defective neurons that are supposed to send signals out to muscles from the brain [18].

Rapid rate transcranial magnetic stimulation has been used to treat patients with migraines, depression, and obsessive-compulsive disorder. Many people suffering from depression respond to antidepressant drugs and psychotherapy, but some do not. Those treatment-resistant patients may benefit from magnetic stimulation. Unlike electroconvulsive therapy, magnetic stimulation does not provoke a seizure, and therefore does not require anesthesia. A patient typically receives magnetic stimulation for half an hour, 5 days a week, for several weeks. The technique has few side effects. Indian psychiatrists Aditya Somani and Sujita Kar have reviewed the evidence for and against using magnetic stimulation in the treatment of depression, including several meta-analyses, and believe it might soon become routine [19].

The Evidence Supporting TENS

Heart pacemakers, neural prostheses, deep brain stimulation, and transcranial magnetic stimulation represent dramatic biomedical advances. Given all these impressive breakthroughs, you might have great expectations for the technique we discussed at the start of this chapter: transcutaneous electrical nerve stimulation (TENS) (Fig. 5.10). Unfortunately, the success of TENS has been uninspiring.

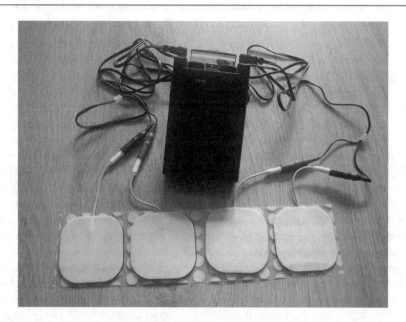

Fig. 5.10 A transcutaneous electrical nerve stimulation (TENS) device. The four electrodes (white) are attached to the skin. (Yeza, CC BY-SA 4.0, https://creativecommons.org/licenses/by-sa/4.0, via Wikimedia Commons)

A challenge faced by TENS is that it must excite sensory nerve fibers that send signals to the brain without exciting motor nerve fibers that tell muscles to contract. Another difficulty is that, in addition to providing the brain with information about pain, sensory fibers also provide information about touch, pressure, and motion. Researchers are still trying to determine the mechanism by which TENS works. One hypothesis, known as "gate theory," is that if you can overwhelm the nervous system with other sensory signals, the body will not be as sensitive to signals for pain. Another hypothesis is that TENS activates opioid receptors, the same molecules affected by drugs such as morphine. Pain is a subjective experience and depends on a patient's emotional and psychological state. Therefore, just because TENS uses electric fields strong enough to stimulate nerves does not necessarily mean it will work as advertised.

A fundamental question in TENS clinical trials is the quality of the placebo. Can a sham TENS treatment provide the same sensation as TENS without stimulating the nerves? Investigators go so far as to build identical-appearing sham devices, complete with blinking lights to indicate the stimulus is on, but which deliver no current. If TENS works by distracting patients so they do not notice their pain, then a placebo should be as effective as the genuine device. If TENS works by stimulating the spinal cord to provide true blockage of pain signals to the brain, then a placebo should have no effect at all.

In 2000, Sarah Milne and her coworkers performed a meta-analysis of randomized, controlled clinical trials for chronic lower back pain. Their study is included in the Cochrane Database of Systematic Reviews [20]. The Cochrane Library is named after Archie Cochrane, a Scottish doctor who advocated for randomized, controlled clinical trials to improve medical care [21]. It collects systematic reviews and meta-analyses about medical research and is an invaluable resource for evidence-based medicine. Milne and her colleagues examined five trials that included a total of 251 patients receiving TENS and 170 patients receiving a sham control. They found no statistically significant differences between the TENS group and the control group. They concluded that "the results of the meta-analysis present no evidence to support the use of TENS in the treatment of chronic low back pain" [20].

More recently, Lien-Chen Wu and his colleagues at the Taipei Medical University in Taiwan conducted a meta-analysis of a larger group of studies that used TENS to treat chronic back pain [22]. They compared 12 randomized, controlled clinical trials that included a total of 700 patients. The controls varied between studies, and included sham treatment, placebos, or treatment with drugs. They concluded that, although TENS might allow the patients to better cope with daily physical activities, it "does not improve symptoms of lower back pain" [22].

Sometimes a major scientific professional organization will sponsor a review of a medical procedure. The American Academy of Neurology, a society representing over 36,000 neurologists and neuroscientists, conducted an evidence-based review and concluded that "transcutaneous electric nerve stimulation… is not recommended for the treatment of chronic low back pain" [23].

Several other meta-analyses have been performed, and the consensus is that the clinical studies to date are either inconclusive or mildly negative about the use of TENS. Harriet Hall [24], a former Air Force medical doctor who now writes about pseudoscience for the website sciencebasedmedicine.org, concluded that "there is some evidence for using TENS for some pain conditions, but that it doesn't appear to be very effective for anything else" and that "it may be worth trying for patients who are desperate, but they should keep their expectations low" [25].

Clyde Norman Shealy, the inventor of transcutaneous electrical nerve stimulation, has become an advocate for alternative medicine. In 1978 he founded the American Holistic Medical Association. He is a proponent of "medical intuition," a form of psychic healing or medical clairvoyance. He also promotes transcutaneous acupuncture, which works by electrical stimulation combined with application of "bliss oils" and makes use of the "five sacred energetic acupuncture rings: air, fire, earth, crystal, and water" [26]. In 2019 Shealy received the Albert Einstein Award of Medicine from the International Association of Who's Who, for his exemplary achievements within the field of Health Care [27]. The Encyclopedia of American Loons describes him as a "remarkably versatile, prolific and respected crackpot, madman and woo-meister" [28].

References

1. Shealy, C. N., Mortimer, J. T., & Reswick, J. B. (1967). Electrical inhibition of pain by stimulation of the dorsal columns: Preliminary clinical report. *Anesthesia & Analgesia, 46*, 489–491.
2. Jeffrey, K. (2001). *Machines in our hearts: The cardiac pacemaker, the implantable defibrillator, and American health care.* Johns Hopkins University Press.

3. Lillehei, C. W., Gott, V. L., Hodges, P. C., Long, D. M., & Bakken, E. E. (1960). Transistor pacemaker for treatment of complete atrioventricular dissociation. *Journal of the American Medical Association, 172*, 2006–2010.

4. Chardack, W. M., Gage, A. A., & Greatbatch, W. (1960). A transistorized, self-contained, implantable pacemaker for the long-term correction of complete heart block. *Surgery, 48*, 643–654.

5. Roth, B. J. (1995). A mathematical model of make and break electrical stimulation of cardiac tissue by a unipolar anode or cathode. *IEEE Transactions on Biomedical Engineering, 42*, 1174–1184.

6. Hodgkin, A. L., & Huxley, A. F. (1952). A quantitative description of membrane current and its application to conduction and excitation in nerve. *Journal of Physiology, 117*, 500–544.

7. https://www.youtube.com/embed/Wd_gKJoo25Y. Access date: January 12, 2022.

8. Baker, P. F., Hodgkin, A. L., & Shaw, T. I. (1962). Replacement of axoplasm of giant nerve fibres with artificial solutions. *Journal of Physiology, 164*, 330–354.

9. Hille, B. (2001). *Ion channels of excitable membranes* (3rd ed.). Sinauer Associates.

10. https://www.youtube.com/watch?v=ZZZsresk840. Access date: January 12, 2022.

11. Ashcroft, F. (2012). *The spark of life: Electricity in the human body*. W. W. Norton.

12. Chase, V. D. (2006). *Shattered nerves: How science is solving modern medicine's most perplexing problem*. Johns Hopkins University Press.

13. https://www.youtube.com/watch?v=QTxuOm_d0dw. Access date: January 12, 2022.

14. Pagnin, D., de Queiroz, V., Pini, S., & Cassano, G. B. (2004). Efficacy of ECT in depression: A meta-analytic review. *Journal of ECT, 20*, 13–20.

15. Catania, K. (2014). The shocking predatory strike of the electric eel. *Science, 346*, 1231–1234.

16. Roth, B. J., Cohen, L. G., & Hallett, M. (1991). The electric field induced during magnetic stimulation. *Electroencephalography and Clinical Neurophysiology, 43*(Supplement), 268–278.

17. Pascual-Leone, A., Dhuna, A., Roth, B. J., Cohen, L., & Hallett, M. (1990). Risk of burns during rapid-rate magnetic stimulation in presence of electrodes. *The Lancet, 336*, 1995–1996.

18. Hallett, M., & Cohen, L. G. (1989). Magnetism: A new method for stimulation of nerve and brain. *Journal of the American Medical Association, 262*, 538–541.

19. Somani, A., & Kar, S. K. (2019). Efficacy of repetitive transcranial magnetic stimulation in treatment-resistant depression: The evidence thus far. *General Psychiatry, 32*, e100074.

20. Milne, S., Welch, V. A., Brosseau, L., Saginur, M. M. D. S., Shea, B., Tugwell, P., & Wells, G. A. (2000). Transcutaneous electrical nerve stimulation (TENS) for chronic low-back pain. *Cochrane Database of Systematics Reviews, 3*, CD003008.

21. Cochrane, A. L. (1972). *Effectiveness and efficiency: Random reflections on health services*. Nuffield Trust.

22. Wu, L.-C., Weng, P.-W., Chen, C.-H., Huang, Y.-Y., Tsuang, Y.-H., & Chiang, C.-J. (2018). Literature review and meta-analysis of transcutaneous electrical nerve stimulation in treating chronic back pain. *Regional Anesthesia & Pain Medicine, 43*, 425–433.

23. Dubinsky, R. M., & Miyasaki, J. (2010). Assessment: Efficacy of transcutaneous electric nerve stimulation in the treatment of pain in neurological disorders (an evidence-based review). *Neurology, 74*, 173–176.

24. Hall, H. (2008). *Women aren't supposed to fly: The memoirs of a female flight surgeon*. iUniverse.

25. https://sciencebasedmedicine.org/tens-for-pain-relief-does-it-work. Access date: January 12, 2022.

26. https://normshealy.com/transcutaneous-acupuncture. Access date: January 12, 2022.

27. https://www.24-7pressrelease.com/press-release/465670/dr-c-norman-shealy-has-been-honored-with-the-albert-einstein-award-of-medicine-by-the-international-association-of-whos-who. Access date: January 12, 2022.

28. https://americanloons.blogspot.com/2012/10/355-c-norman-shealy.html. Access date: January 12, 2022.

Is Your Cell Phone Killing You?

6

In her book *Disconnect*, American epidemiologist and toxicologist Devra Davis writes that "cell phones have revolutionized the world… But is it possible that the pervasive use of cell phones is causing a host of subtle, chronic health problems today, damaging our ability to have healthy children and creating long-term risks to our brains and bodies?" [1] (Fig. 6.1). Davis is the founder of the Environmental Health Trust [2], a non-profit organization that contends radio-frequency radiation, such as that associated with cell phones, endangers our health. She has been called a crusader in the fight over cell phone safety [3].

Davis's claims are examined in this chapter. Our goal is to determine whether or not her allegations are supported by the evidence. Does your iPhone pose a health hazard, or will her assertions that we are all at risk be proven baseless with additional study? Is there really a disconnect between what scientists know about the risks of radio frequency radiation and the public's widespread acceptance of cell phones as harmless?

Maxwell's Equations

To understand electromagnetic radiation associated with cell phones, we must learn more physics. In Chap. 2 we found that electric currents (moving charges) produce magnetic fields. Then Chap. 3 taught us that charges, even if not moving, create electric fields. Next, Chaps. 4 and 5 explained how a changing magnetic field induces an electric field: electromagnetic induction. This chapter supplies the last piece of the puzzle: A changing electric field produces a magnetic field. This effect was discovered by Scottish physicist James Maxwell in the 1860s (Fig. 6.2). Maxwell built on the work of Michael Faraday, the discoverer of electromagnetic induction who was a brilliant experimentalist but was poor at math. In contrast, Maxwell—who was Second Wrangler in the famed Cambridge University Mathematical Tripos Exam—was excellent at math and expressed his ideas about

© The Author(s), under exclusive license to Springer Nature Switzerland AG 2022
B. J. Roth, *Are Electromagnetic Fields Making Me Ill?*,
https://doi.org/10.1007/978-3-030-98774-9_6

Fig. 6.1 A person talking on a cell phone. (From https://www.fda.gov/ radiation-emitting-products/cell-phones/ do-cell-phones-pose-health-hazard)

Fig. 6.2 James Clerk Maxwell (1831–1879). (George J. Stodart, Public domain, via Wikimedia Commons)

charge, current, electric fields, and magnetic fields in what are today known as Maxwell's equations. This remarkable set of four partial differential equations are so honored among physicists that they sometimes wear them on their tee shirt (Fig. 6.3).

Using his equations, Maxwell made one of the most famous predictions in physics: A changing magnetic field induces a changing electric field, and this changing electric field produces a changing magnetic field, which creates a changing electric field, which causes a changing magnetic field, and so on. The result is an electromagnetic wave that travels through space as radiation (Fig. 6.4). In addition, Maxwell was able to predict how fast these waves propagate: at the speed of light. Visible light is one type of electromagnetic radiation.

Fig. 6.3 Maxwell's equations on a tee shirt. The electric field is **E**, the magnetic field is **B**, the current density is **J**, and the charge density if ρ. The symbols ε_0 and μ_0 are constants, and $\nabla\bullet$, $\nabla\times$, and $\partial/\partial t$ are mathematical operators used in calculus

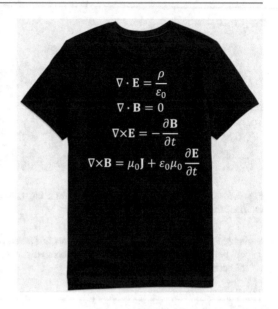

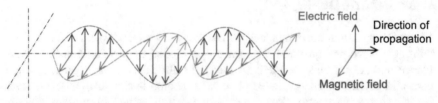

Fig. 6.4 An electromagnetic wave. The electric (blue) and magnetic (red) fields oscillate in space and point perpendicular to each other. The entire distribution propagates to the right, so that at a particular location the electric and magnetic fields oscillate in time. (SuperManu, CC BY-SA 3.0, http://creativecommons.org/licenses/by-sa/3.0/, via Wikimedia Commons)

The light we see is by no means the only type of electromagnetic wave. Electric and magnetic fields can oscillate at any frequency (number of oscillations per second). Radio waves typically oscillate millions or billions of times per second (with frequencies measured in megahertz or gigahertz). Infrared light oscillates trillions of times per second. Even higher frequencies correspond to visible light, ultraviolet light, x-rays, and gamma rays. The electromagnetic spectrum suddenly made sense when Maxwell discovered his equations (Fig. 6.5).

How influential was Maxwell's discovery? It brought electricity, magnetism, and optics together into a single unified theory. One of the most essential classes any undergraduate physics major takes is a course in electromagnetism, which is basically a semester-long study of Maxwell's equations. American physicist Richard Feynman writes that:

from a long view of the history of mankind—seen from, say, ten thousand years from now—there can be little doubt that the most significant event of the 19th century will be

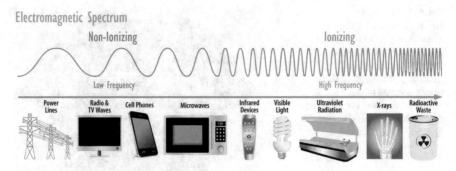

Fig. 6.5 The electromagnetic spectrum. (From: https://www.niehs.nih.gov/health/topics/agents/emf/index.cfm)

judged as Maxwell's discovery of the laws of electrodynamics. The American Civil War will pale into provincial insignificance in comparison with this important scientific event of the same decade. [4]

Wave/Particle Duality

Maxwell's equations are not, in fact, the last piece of the puzzle. Maxwell described light as a wave, but quantum mechanics—developed by physicists such as Max Planck and Albert Einstein—tells us that light also behaves as a particle. Light comes in packets of energy called photons. A photon is similar to an elementary particle such as an electron; you can have one photon, or two, or millions, but you cannot have half a photon. Maxwell's equations do not predict photons. These quanta were discovered in the twentieth century and are part of modern physics. Having two complementary views of light is known as wave/particle duality. Both views are correct; sometimes light acts like a wave, and sometimes it acts like a particle. Wave/particle duality is one of the peculiar predictions of quantum mechanics.

The frequency of an electromagnetic wave determines a photon's energy. A 1 MHz radio wave has a photon energy of only 0.000000004 electron volts. Many cell phones operate at a frequency around 1 GHz, with the corresponding photon energy of 0.000004 electron volts. The new 5G cell phones will make use of frequencies up to nearly 1000 GHz, with a photon energy of 0.004 electron volts. What should we compare these energies to? We saw in Chap. 2 that molecules move about in random motion, and at room temperature they have a thermal energy of about 0.02 electron volts. Atoms are held together in molecules by chemical bonds, and even the weakest of these bonds are typically several electron volts. To produce photon energies big enough to ionize atoms, break chemical bonds, or form free radicals (a highly reactive molecule that can damage DNA), you need ultraviolet light or x-rays. For instance, a typical x-ray in your dentist's office has an energy of about 50,000 electron volts, which is sufficient to rip an electron right out of an atom.

Because a single photon of high frequency electromagnetic radiation, such as an x-ray, has enough energy to remove an electron from a molecule, leaving behind a charged ion, it is known as ionizing radiation. If an x-ray photon hits a DNA molecule, it can change its molecular structure. Such damage to DNA might cause a mutation; sometimes it turns a normal cell into one that grows out of control, forming a tumor. That is why ionizing radiation like x-rays can lead to cancer. Even an ultraviolet photon, which has less energy than an x-ray photon, has enough energy to produce skin cancer.

The electromagnetic radiation used by cell phones has a photon energy that is at most a few thousandths of an electron volt and usually much less. This is not enough energy to rupture chemical bonds and create ions; it is nonionizing. It is even less than the thermal energy that molecules have as they jostle about. Cell phone radiation cannot damage a DNA molecule, thereby causing cancer. Does this mean cell phones have nothing to do with cancer? Although they cannot cause cancer directly, they may lead to cancer indirectly. Our bodies have several physiological mechanisms that determine how susceptible we are to tumor growth, such as our immune system. Furthermore, our cells have ways to repair some of the damage to DNA that was produced by, say, an x-ray photon. If cell phone radiation is able to influence these physiological mechanisms, it could encourage cancer growth indirectly. Therefore, to understand the relationship between cell phones and cancer, we need to search for such indirect biological effects.

Even though a single photon might not have enough energy to disrupt a molecule, if you have an intense beam containing lots of photons then maybe they could together break chemical bonds. Scientists have a name for what happens when many photons dump lots of energy into tissue: heat. The study of heat and temperature is known as thermodynamics. This field of physics was developed in the nineteenth century, at about the same time Maxwell was deriving his equations of electromagnetism. The laws of thermodynamics are among the most fundamental principles of science. We have a good understanding of thermodynamics even for a complex organism, which contains feedback loops aimed at holding its body temperature constant.

The intensity of cell phone radiation is not sufficient to warm our body much. Cell phones do not cause cancer by locally heating our brain. We can determine the temperature of our brain when exposed to cell phone radiation, and it does not increase enough to make a difference. If you want to explain how cell phones cause cancer, you should look for a mechanism that is not thermal, but instead is nonthermal, meaning that it does not depend on increasing the temperature. We know electric and magnetic fields are capable of causing biological responses by nonthermal mechanisms. After all, stimulation of an action potential in a nerve axon has nothing to do with raising the temperature of the axon. But nerve excitation occurs for electric field pulses lasting on the order of a millisecond, or in other words for frequencies on the order of a kilohertz. Are there analogous nonthermal mechanisms for radio frequency waves? The rest of this chapter attempts to answer that question.

To summarize, our analysis of the physics of electromagnetic radiation highlights two critical facts:

1. A photon from a cell phone is so weak that it cannot break bonds in a DNA molecule and cause cancer directly. If cell phones cause cancer the effect must be indirect.
2. Cell phone radiation is not intense enough to heat tissue. Any mechanism by which cell phone radiation causes cancer must be nonthermal.

Electromagnetic Radiation and Heat

The subject of heating tissue with electromagnetic radiation is so important that we should investigate it in more detail.

Everyone knows you can heat tissue with electromagnetic fields; we do it every time we use our microwave oven (Fig. 6.6). Microwaves are electromagnetic waves with a frequency from 0.3 to 300 GHz. Think of microwaves as occupying the transition zone on the electromagnetic spectrum between radio waves and infrared light (Fig. 6.5). There is no doubt that microwaves, if strong enough, could cook your brain. Are microwaves from cell phones that strong? Not according to Eleanor Adair (Fig. 6.7).

Adair was an American physiologist who performed the first studies examining the safety of microwave radiation. After receiving an undergraduate degree from Mount Holyoke College, Adair obtained her doctorate from the University of Wisconsin, where her interdisciplinary studies combined the two fields of psychology and physics. She worked at the John B. Pierce Laboratory in New Haven, Connecticut—a non-profit organization affiliated with Yale University—and was married to Robert Adair, the Yale physicist we discussed in Chap. 4. Her research focused on how our bodies regulate their temperature when exposed to microwaves. She studied microwave radiation up to eight times as strong as current exposure standards and compared continuous and pulsed waves. She helped develop safety standards for microwave radiation by chairing committees formed by the Institute

Fig. 6.6 A microwave oven. (Consumer Reports, CC BY-SA 4.0, https://creativecommons.org/licenses/by-sa/4.0, via Wikimedia Commons)

Fig. 6.7 Eleanor Adair
(1926–2013). (From
Murphy, M. R. (2003)
Bioelectromagnetics,
24:S1–S2. Photo: Public
Domain)

of Electrical and Electronics Engineers (commonly known as the IEEE and pronounced "I triple E"). In 2007 she was awarded the prestigious d'Arsonval Award by the Bioelectromagnetics Society.

The only biological responses to microwaves that Adair found in her experiments were related to tissue heating. She advocated for uniform exposure standards based on thermal heating as "one step forward toward mitigating many traditional concerns [about radio frequency radiation]—e.g., the science is inconclusive, all radiation is hazardous and should be avoided (no level of exposure is safe), artificial sources of anything are more dangerous than natural sources—and opening the door for the acceptance of innovative and beneficial technologies" [5].

How much heating is safe? When considering high-frequency electromagnetic radiation, researchers do not take the electric or magnetic field strength as the key parameter. Instead, they use the specific absorption rate: the rate of energy absorbed by the tissue per unit mass, measured in watts per kilogram (W/kg). A watt is a unit of power indicating the rate of energy use (a watt is a joule of energy per second), and a kilogram is the metric unit of mass. To give you some perspective, the human body generates about 100 watts of metabolic energy when at rest, and a typical person has a mass of about 70 kilograms, implying a heat production of 1.4 W/kg.

The current guideline for the maximum specific absorption rate produced by radiation from a cell phone, established by the Federal Communications Commission (FCC), is 1.6 W/kg. This is an average value over time. If the cell phone was emitting radiation continuously, then 1.6 W/kg is the maximum power it is allowed to deposit in the brain. If the radiation came in pulses so that 99% of the time there is no radiation at all, then during the 1% of the time when the radiation is present the specific absorption rate could be as high as 160 W/kg and would still meet the FCC standard. On the other hand, the specific absorption rate cannot rise above 1.6 W/kg anywhere in the tissue. If the radiation is restricted to a region of the brain right under your ear, the specific absorption rate must be kept below 1.6 W/kg everywhere, including just below your ear. In other words, you are not allowed to make up for a higher absorption rate at one location by having a lower rate at another.

How does a specific absorption rate translate into an increase in temperature? University of Pennsylvania engineer Kenneth Foster and his coworkers have developed a mathematical model that predicts the temperature increase from the specific absorption rate and a few tissue parameters [6]. If 1 GHz microwave radiation is incident on the head so that at the brain surface the specific absorption rate is 1.6 W/kg, then the steady-state temperature increase at the surface is about 0.05 °C, equivalent to 0.09 °F, which is a factor of 10 less than the normal variation of body temperature throughout the day. The calculation includes both heat conduction in the brain tissue and heat convection by blood flow.

Previous chapters in this book focused on the strength of the electric field. How does this field strength relate to the specific absorption rate? Foster has calculated that, for a continuous electromagnetic wave with a frequency of 1 GHz, a specific absorption rate of 1 W/kg is equivalent to an electric field of 30 V/m [7]. The associated magnetic field would have a strength of about 0.1 μT, which is so weak that any interactions between an electromagnetic wave and tissue are believed to act through the electric field.

The threshold for activation of a neuron is about 10 V/m. Can a 30 V/m microwave then trigger neural excitation? No. The reason this is not possible is because neurons have their own characteristic time of about 1 millisecond. If the time it takes for the electric field to oscillate is much shorter than 1 millisecond, the neuron cannot respond. An engineer would say the nerve acts as a low-pass filter, removing the high frequencies. A 1 GHz microwave oscillates back and forth once in 0.000001 milliseconds. You cannot excite nerves or muscle with such high frequency fields.

Our analysis indicates that a specific absorption rate below the maximum rate allowed by the FCC results in negligible tissue heating. The thermal mechanism underlying how weak electric fields from a cell phone interact with the brain cannot cause tissue damage. Before analyzing potential nonthermal mechanisms, let us take a brief detour and describe three medical applications of electromagnetic fields that either heat tissue or have an effect because the electric field is so strong.

Diathermy

Radio waves and microwaves are sometimes used to heat tissue as a form of therapy, called diathermy [8]. Shortwave diathermy makes use of radio frequencies, such as 27 MHz, while microwave diathermy usually employs a frequency similar to that found in a microwave oven (2500 MHz). Shortwave diathermy heats deep tissue without burning the skin. Microwave diathermy penetrates less deeply, and is adopted especially for superficial heating. A specific absorption rate of around 20 W/kg (corresponding to an electric field of about 200 V/m) will heat tissue several degrees Celsius. Heating improves blood flow, helping tissues heal. This technique is used to treat sports injuries and for physical therapy.

Radio Frequency Ablation

What happens if you deliver a huge amount of electrical energy into tissue? If the amount of energy is large enough, the temperature can rise so much that the tissue vaporizes. This technique, called radio frequency ablation, is used to treat an irregular heartbeat known as atrial fibrillation, during which action potentials propagate in a chaotic manner [9]. The cardiologist inserts a catheter electrode through a vein into the heart and removes (ablates) small regions in the atria, breaking up the path over which action potentials propagate, thereby helping the heart to maintain a normal rhythm. The electric field oscillates with a radio frequency of about 0.5 MHz and is strong (10,000 V/m near the electrode tip) resulting in a specific absorption rate of about 25,000 W/kg. This is enough to bring the temperature near the tip to approximately the boiling point, 100 °C.

Electroporation

If a giant electric field is applied for only an instant, it may not have time to heat up the tissue. Such a strong field can, nevertheless, affect a cell by creating holes in the cell membrane in a process called electroporation [10]. A 50,000 V/m electric field lasting for a few microseconds induces a voltage across the cell membrane of about three quarters of a volt. This causes pores to form in the membrane with a radius of around 0.001 microns, which then last for up to several seconds after the pulse is turned off. These pores are larger than those in an ion channel and are therefore nonspecific, letting in not only ions but also moderately sized molecules. If too powerful, the shock kills the cell. In other cases, the cell recovers but now contains large molecules previously restricted to the extracellular space. One use of electroporation is to deliver drugs or even modified DNA into a cell. In addition, electroporation is being explored as an alternative to radio frequency ablation for destroying tissue in the heart when treating atrial fibrillation.

Indirect and Nonthermal Mechanisms

If any mechanisms exist for cell phone radiation-induced cancer (or other health threats), then those mechanisms must be indirect and nonthermal. Physicist Asher Sheppard, of Loma Linda University in California, and his coworkers have reviewed numerous potential mechanisms in an article that was published in the journal *Health Physics* [11]. We discuss a few of these mechanisms below:

1. An ion takes about 0.1 microseconds to cross a membrane. That means for frequencies above 10 MHz the ion would not have time to be pushed across the membrane before the electric field changed direction and pushed it back. Cell phone radiation at a frequency of 1 GHz would oscillate back and forth one hun-

    dred times faster than an ion could traverse the membrane. This makes mechanisms for ion transport unlikely at cell phone frequencies.

2. Earlier we said that the interaction of many photons with tissue is a form of heat. This statement is not strictly true. If the photons all act independently, they create heat. But if they act in unison (coherently) they could potentially excite some molecules to new energy states. Such a multiphoton process is more common for strong electromagnetic fields. Sheppard et al. calculate that even for radiation strong enough that the specific absorption rate is 100 W/kg, the probability of an observable multiphoton process is "indistinguishable from zero" [11].

3. Some researchers have postulated that radio frequency energy may be deposited preferentially in tiny structures, so that you get a significant temperature increase in a local region, but the temperature increase averaged over the entire tissue remains negligible. However, the time needed for heat to diffuse over, say, the length of a cell is so short (about a millisecond) that any such temperature gradients would immediately vanish.

4. In Chap. 4, we discussed Adair's analysis of resonance absorption of energy for very low frequency electromagnetic radiation. Sheppard et al. examine analogous resonance absorption at radio frequencies [11]. For instance, large macromolecules vibrate at discrete frequencies. If an electric field oscillates at the same frequency, then the two would be in resonance, allowing the field to pump energy into the resonant mode. (A non-biological example would be a wine glass that, when you tap it, rings at a specific frequency; if a singer belts out the same pitch, energy can be pumped into the resonance mode until the glass shatters.) When Sheppard et al. reviewed known resonant modes of molecules, taking into account the extreme damping of these modes by the frictional properties of water, they found that resonant absorption is unlikely for microwave frequencies less than 300 GHz.

5. For a linear process, if you double the electric field strength you double its effect. Also, if you change the sign of the electric field (so the field now points in the opposite direction) you change the sign of its effect. When an electric field is oscillating at a high frequency, any linear response caused by the field in one direction will be undone when the field is in the opposite direction. In order to get a net response, the process must be nonlinear: it must be larger for one direction of the electric field than for the other. Mechanisms for how cell phone radiation interacts with tissue must be not only indirect and nonthermal, but also nonlinear. Such nonlinear effects exist throughout biology (excitation of an action potential is an example), but they generally require strong electric fields.

6. One nonlinear phenomenon is rectification. If the response to an electromagnetic wave is rectified, then when the electric field points in one direction there is an effect, while in the opposite direction there is not. This would mean that the response to a high-frequency wave would not average to zero. If the wave's amplitude varied, perhaps because the radiation is applied in pulses, rectification could change a high-frequency wave into a lower-frequency signal (a process called demodulation), which would be more likely to have a biological response. Of particular interest would be a demodulated signal acting at the cell mem-

brane. Several researchers have proposed such mechanisms. Sheppard et al. analyzed these effects and concluded that "fundamental considerations show that for any practical exposure the demodulated [extremely low frequency] signal that may exist across the membrane would be irretrievably lost in membrane noise" [11].

7. Herbert Fröhlich, a German-born British physicist, has postulated that a high-frequency electromagnetic field could excite many molecules to vibrate together so that they all oscillate back and forth coherently. His analysis suggested that energy pumped into such oscillations would be concentrated into the lowest frequency vibrational mode, a process known as Fröhlich condensation. Sheppard et al. analyzed this behavior, and concluded that "coherent mechanisms in biological cells would require coupling among many, if not all, molecules at the membrane surface, but still could not avoid destruction by damping forces, nor last long enough to play any role for a… [frequency] below about [1000–10,000 GHz]."

Sheppard et al. concluded that "in the frequency range from several megahertz to a few hundred gigahertz, the focus of this paper, the principal mechanism for biological effects, and the only well-established mechanism, is the heating of tissues" [11]. Their conclusions are shared by Eleanor Adair. When addressing the question of nonthermal mechanisms for tissue damage by radio frequency waves, Adair wrote that "the recent controversial claims of nonthermal effects, especially [radio frequency/microwave] exposures that are amplitude modulated at [extremely low] frequencies, have neither been substantiated experimentally, nor replicated independently" [5].

Theoretical analysis appears to rule out any biological response to high frequency radiation. But what does the data show? Is there any evidence from laboratory experiments or epidemiology studies to support the contention that cell phones cause cancer? Let us start answering these questions by looking at brain cancer trends.

Brain Cancer Trends

In Chap. 4 we saw that leukemia incidence had not increased throughout the twentieth century as the United States adopted electricity, suggesting that power lines do not cause cancer. In this chapter, we examine brain cancer incidence over the past few decades as cellular phone use has exploded. Peter Inskip, an epidemiologist with the National Cancer Institute, and his coworkers found that the incidence of brain cancer between 1992 and 2006 was generally flat or trended slightly downward (Fig. 6.8) [12]. The one group that showed an increasing trend—20–29 year-old women—had additional cancers in the frontal lobe, not the temporal lobe. (The temporal lobe is the part of the brain under the ear, and the most likely place for a cell phone to cause cancer.) According to statistics from the National Cancer Institute, this flat or downward trend holds for children and for people in other countries [13]. With cell phone use increasing dramatically over the last few decades,

Fig. 6.8 Cell phone use and brain cancer trends between 1976 and 2006. (Data from Inskip et al. [12])

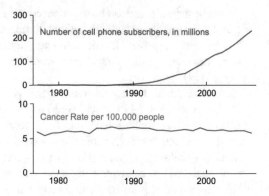

why did the brain cancer rate not spike? One possibility is that cancer takes a long time to develop, and we will see a surge in the future. At the moment, we cannot know for sure whether or not this hypothesis is correct, but it becomes less likely as each year goes by with no rise in the incidence of brain cancer.

Point/Counterpoint Debate

In Chap. 2, we found that point/counterpoint articles in the journal *Medical Physics* were a valuable teaching tool and provided insight into medical controversies. The December 2008 point/counterpoint article debated the proposition that "there is currently enough evidence and technology available to warrant taking immediate steps to reduce exposure of consumers to cell-phone-related electromagnetic radiation" [14]. Arguing for the proposition was Vini Khurana, of the Australian National University in Canberra. Khurana asserted on the Larry King Live television show that cell phones are more dangerous than smoking. Arguing against the proposition was John Moulder, of the Medical College of Wisconsin, who we met during our discussion of power lines and cancer.

Khurana began by citing his own meta-analysis of epidemiological data, which analyzed the INTERPHONE study administered by the World Health Organization and the studies of Swedish oncologist Lennart Hardell [15]. The Swedish studies found a significant link between cell phones and brain cancer. The results of the larger INTERPHONE study were mixed, with no risk for most users but a suggestion of increased risk among the subjects who talked on their cell phones most. Critics of the INTERPHONE study say that it did not reflect the current heavy use of cell phones. During this study (2000–2006) cell phones were not as ubiquitous as they are now, and the 10% of subjects who talked on their phone the most used it for only about 30 minutes per day. Khurana then discussed the BioInitiative Report, a 2007 summary by a group of 14 scientists who assessed the evidence supporting a relationship between cell phones and cancer, which determined that additional safety regulations are warranted [16]. Khurana concluded "there are reasonable grounds to suspect a looming public health tragedy" [14].

In Moulder's opening statement, he referred to his own review of the scientific literature, which determined that the evidence for an association between cancer and radio frequency energy is unconvincing [17]. He then asserted that no biophysical mechanisms are known that could explain such biological behavior, and that some of the early experimental evidence of nonthermal mechanisms is either irreproducible or flawed. Moulder concluded that "weak epidemiological evidence of an association of mobile phone use with brain cancer incidence, when combined with the biophysical implausibility of a causal link and the strongly unsupportive animal studies, does not support the case that regulation of mobile phone use is urgently needed" [14].

In their rebuttals, Khurana and Moulder each argued that his evidence is better than the other guy's. Moulder noted that the BioInitiative Report, which Khurana cited, was not peer reviewed. During peer review an article or grant proposal is sent to leading experts in the field. These reviewers have two duties: (1) give a yes or no recommendation about publishing the article or funding the proposal, and (2) make suggestions about how the article or proposal might be improved. Often reviewers' comments lead to revisions, followed by additional review. This process is sometimes painful for authors, but it ensures the quality of the research. Peer review is one of the pillars on which modern science is built. A lack of peer review is a not a minor point. It is a major concern.

Khurana and Moulder's point/counterpoint article summarized the case for and against a link between cell phones and cancer. It also illustrated how the evidence on this topic is so diverse that a reasonable-sounding argument can be made to support either position.

Epidemiological Studies

The INTERPHONE epidemiological study, referred to by both Devra Davis in *Disconnect* and by Vini Khurana in the point/counterpoint article, was a crucial piece of evidence in the cell phone/cancer debate. INTERPHONE was a *case-control* study in which researchers analyzed patients suffering from cancer by examining their past cell phone use retrospectively. Two more epidemiological studies, one performed in Denmark and the other in Great Britain, were *cohort* studies in which a large group of initially healthy people were followed to determine who got sick and who did not [13]. Case-control studies are quick, easy, and useful for rare diseases or those with a long latency period, but they are prone to bias. Cohort studies take longer and cost more, but they are better for determining the risk to a population. Neither case-control studies nor cohort studies are as reliable as randomized, placebo-controlled, double-blind clinical trials.

The Danish trial was a nationwide cohort study, analyzing all Danes over 30 years old. It divided the population into mobile phone subscribers and non-subscribers, and followed these groups for almost 20 years. The study found no increased risk of brain tumors among cell phone subscribers. One weakness of this study is that it analyzed the risk from cell phone subscription, not cell phone use directly.

The Million Women Study followed a cohort of almost 800,000 middle-aged women in Great Britain for 10 years. It analyzed cell phone use, not merely subscription, and concluded that mobile phones were not associated with an increased incidence of cancer.

In 2019, Martin Röösli, of the University of Basil in Switzerland, and a team of European investigators performed a meta-analysis of these epidemiological studies and others [18]. In their publication, they point out one difficulty in determining cell phone radiation exposure. The recent use of adaptive power control technology has lowered cell phone radiation emission so that exposure may be less in the last few years compared to a decade ago, even if the number of hours of cell phone use has increased. Adaptive control was invented to reduce the power of the emitted cell phone signal whenever possible, with the dual goals of increasing battery life and preventing interference between adjacent cellular networks. Ironically, a reduction of human exposure to microwave radiation might be an unintended byproduct of this technology. This observation highlights how difficult it is to assess cell phone exposure accurately.

Röösli et al. discuss the relative strengths and weaknesses of case-control versus cohort studies [18]. They mention that a large cohort study, called COSMOS, is now following approximately 100,000 volunteers in the United Kingdom and 200,000 in Europe. COSMOS began in 2010, and participants will be followed for 20 years. Each participant will complete an online questionnaire probing their health, lifestyle and cell phone use. The INTERPHONE study was susceptible to recall bias, in which cancer victims tend to remember talking on the phone more than controls with no cancer, even if their phone use was the same. COSMOS will avoid recall bias by obtaining cell phone records from mobile phone companies to supplement the questionnaire.

After discussing all these topics and analyzing numerous clinical trials, Röösli et al. concluded that "epidemiological studies do not suggest increased brain or salivary gland tumor risk with [mobile phone] use, although some uncertainty remains regarding long latency periods (>15 years), rare brain tumor subtypes, and [mobile phone] usage during childhood" [18].

Neither epidemiological data, nor cancer incidence statistics, nor theoretical analyses of mechanisms offer much support for a link between cell phones and cancer. Before coming to any decision, however, we should examine another source of evidence: laboratory experiments.

Experimental Studies

In *Disconnect*, Davis described the research of Henry Lai, a bioengineering professor at the University of Washington. Lai and his collaborator Narendra Singh observed damage to DNA, the material that contains our genetic code, after brain cells from rats were exposed to microwaves. The disruption of DNA could lead directly to cancer. However, the review article by Moulder and his colleagues noted that other investigators have been unable to reproduce these results and suggested various possible artifacts that may explain Lai's observations [17].

Davis also highlighted the work of Leif Salford, of Lund University in Sweden. Salford and his colleagues reported experiments suggesting that radio frequency radiation breaks down the blood brain barrier. This selective layer of cells prevents many molecules circulating in the blood from entering the brain. Disruption of the blood brain barrier would have dire consequences for our health. Of particular concern was that Salford et al. found these effects for a specific absorption rate hundreds of times less than the maximum limit permitted for a cell phone.

The blood brain barrier is so essential for our health that many researchers have followed up these initial studies. The results vary, and you can find laboratory experiments supporting both sides of the question. French scientist Anne Perrin and her colleagues critically reviewed the published scientific literature [19]. They examined 16 articles published between 2005 and 2009, including some that set out to reproduce Salford's earlier results. Perrin et al. concluded that the articles they reviewed "provide no convincing proof of deleterious effects of [radio frequency radiation] on the integrity of the [blood brain barrier], for specific absorption rates (SAR) up to 6 W/kg" [19]. Critical reviews like this one examine the experiments in excruciating detail, assessing their strengths and weaknesses. The reviewers almost always bemoan the uneven quality of the research and the inconsistency of the results. The main conclusion often is that the evidence is ambiguous.

Davis herself admitted that the literature is mixed. She wrote "there is no denying that the majority of published studies on radio frequency radiation and the brain do not show any impact. How is this possible?" [1] She then quoted Allan Frey, who performed some of the earliest experiments studying how microwave radiation affects the blood brain barrier. Frey complained that studies meant to replicate the original results often change the methods, which makes them come to different conclusions. While this objection may be true, usually the modifications are meant to improve the experiment, eliminating artifacts and systematic errors.

Davis believed that the research into radio frequency effects was reaching a tipping point, stating that "a tipping point arrives when the way things have appeared to work no longer makes sense. I believe that in the scientific world, we have arrived at that tipping point" [1]. We do seem to be approaching a tipping point in our understanding of how radio frequency radiation affects our health, but the consensus might not be tipping in the direction that Davis expected.

We will consider additional laboratory experiments related to cell phone radiation in the next chapter. To finish this chapter, let us review the opinions of three government agencies.

Recommendations of Government Agencies

The Food and Drug Administration (FDA) is the federal agency responsible for regulating the nation's food, tobacco products, prescription drugs, medical devices, and electromagnetic radiation. In 2020 it published its own review of the scientific literature between 2008 and 2018 with relevance to radio frequency radiation and cancer [20].

One of the main topics of the FDA review was a series of experiments published in 2017 and carried out by the National Toxicology Program, another US government agency. Mice and rats were irradiated with moderately strong (typically 6 W/kg) radio frequency waves. The advantage of animal studies over human clinical trials is that the researchers have more control over the experiment. The disadvantage of working with animals is that the relevance to humans is unclear.

The results of the study were widely discussed in the press because the rats exposed to radiation had a significantly greater risk for heart tumors than the control rats. The results were confusing, however, for two reasons: (1) only male rats showed this correlation, but not females, and (2) despite the tumors, exposed male rats lived longer than the unexposed ones. In addition, the FDA review suggested that the older male rats might have had a more difficult time regulating their body temperature, leading to heating. This would mean that the National Toxicology Program study was looking at thermal effects, reducing its value for assessing non-thermal mechanisms.

In an article appearing in the *Skeptical Inquirer*, Christopher Labos and Kenneth Foster critique the National Toxicology Program study:

> The inherent weakness of the [National Toxicology Program] results is their lack of consistency. We see a signal for harm in rats but not mice. We see a signal for harm in male rats but not female rats. We see a signal for schwannomas [a type of tumor] in the heart but not the rest of the body. Finally, the rats exposed to [radio frequency radiation] actually lived longer on average than the controls. So do cell phones cause cancer while simultaneously extending survival? It is not impossible that there is some yet to be fully understood mechanism at play, but at this point random chance seems far more likely. [21]

The FDA review went on to analyze 39 publications that met its criteria to be part of the study. It concluded that "these studies have not produced any clear evidence that [radio frequency radiation] exposure has any [tumor-causing] effect" [20].

Reviews such as these are a key reason the major health agencies do not believe that cell phones cause cancer. When agency scientists systematically weigh all the evidence, they consistently find no effect. The Centers for Disease Control and Prevention (CDC)—the US government federal agency that is responsible for protecting public health—is a bit more equivocal: "At this time we do not have the science to link health problems to cell phone use" [22]. The National Cancer Institute (NCI) is part of the US National Institutes of Health and is the primary federal agency for cancer research. Many of the nation's best and brightest scientists and doctors work for, or are funded by, the NCI. Anyone who wants expert information about cancer should consult the NCI. On its website, it concludes that "the only consistently recognized biological effect of radiofrequency radiation absorption in humans that the general public might encounter is heating to the area of the body where a cell phone is held (e.g., the ear and head). However, that heating is not sufficient to measurably increase body temperature. There are no other clearly established dangerous health effects on the human body from radiofrequency radiation" [13].

The conclusions of the FDA, CDC, and NCI seem definitive. There is no longer a disconnect between what scientists know and the public's widespread acceptance of cell phones. The issue seemed settled. Then along came 5G.

References

1. Davis, D. (2010). *Disconnect: The truth about cell phone radiation, what the industry has done to hide it, and how to protect your family*. Dutton.
2. https://ehtrust.org. Access date: January 12, 2022.
3. https://www.cnet.com/tech/mobile/cell-phone-radiation-harmless-or-health-risk. Access date: January 12, 2022.
4. Feynman, R. P., Leighton, R. B., & Sands, M. (1964). *The Feynman lectures on physics* (Vol. 2). Addison-Wesley.
5. Adair, E. R. (2002). Biological effects of radio-frequency/microwave radiation. *IEEE Transactions on Microwave Theory and Techniques, 50*, 953–962.
6. Foster, K. R., Lozano-Nieto, A., & Riu, P. J. (1998). Heating of tissues by microwaves: A model analysis. *Bioelectromagnetics, 19*, 420–428.
7. Foster, K. R. (2000). Thermal and nonthermal mechanisms of interaction of radio-frequency energy with biological systems. *IEEE Transactions on Plasma Science, 28*, 15–23.
8. Giombini, A., Giovannini, V., Di Cesare, A., Pacetti, P., Ichinoseki-Sekine, N., Shiraishi, M., Naito, H., & Maffulli, N. (2007). Hyperthermia induced by microwave diathermy in the management of muscle and tendon injuries. *British Medical Bulletin, 83*, 379–396.
9. Joseph, J. P., & Rajappan, K. (2012). Radiofrequency ablation of cardiac arrhythmias: Past, present and future. *QJM: An International Journal of Medicine, 105*, 303–314.
10. Weaver, J. C., Langer, R., & Potts, R. O. (1995). Tissue electroporation for localized drug delivery. In M. Blank (Ed.), *Electromagnetic fields: Biological interactions and mechanisms* (pp. 301–316). American Chemical Society.
11. Sheppard, A. R., Swicord, M. L., & Balzano, Q. (2008). Quantitative evaluations of mechanisms of radiofrequency interactions with biological molecules and processes. *Health Physics, 95*, 365–396.
12. Inskip, P. D., Hoover, R. N., & Devesa, S. S. (2010). Brain cancer incidence trends in relation to cellular telephone use in the United States. *Neuro-Oncology, 12*, 1147–1151.
13. https://www.cancer.gov/about-cancer/causes-prevention/risk/radiation/cell-phones-fact-sheet. Access date: January 12, 2022.
14. Khurana, V. G., Moulder, J. E., & Orton, C. G. (2008). There is currently enough evidence and technology available to warrant taking immediate steps to reduce exposure of consumers to cell-phone-related electromagnetic radiation. *Medical Physics, 35*, 5203–5206.
15. Khurana, V. G., Teo, C., Kundi, M., Hardell, L., & Carlberg, M. (2009). Cell phones and brain tumors: A review including the long-term epidemiologic data. *Surgical Neurology, 72*, 205–215.
16. https://bioinitiative.org. Access date: January 12, 2022.
17. Moulder, J. E., Foster, K. R., Erdreich, L. S., & McNamee, J. P. (2005). Mobile phones, mobile phone base stations and cancer: A review. *International Journal of Radiation Biology, 81*, 189–203.
18. Röösli, M., Lagorio, S., Schoemaker, M. J., Schüz, J., & Feychting, M. (2019). Brain and salivary gland tumors and mobile phone use: Evaluating the evidence from various epidemiological study designs. *Annual Review of Public Health, 40*, 221–238.
19. Perrin, A., Cretallaz, C., Collin, A., Amourette, C., & Yardin, C. (2010). Effects of radiofrequency field on the blood-brain barrier: A systematic review from 2005 to 2009. *Comptes Rendus Physique, 11*, 602–612.
20. Food and Drug Administration. (2020). *Review of published literature between 2008 and 2018 of relevance to radiofrequency radiation and cancer*. FDA.
21. https://skepticalinquirer.org/2018/07/cell-phone-radiation-and-cancer. Access date: January 12, 2022.
22. https://www.cdc.gov/nceh/radiation/cell_phones._faq.html. Access date: January 12, 2022.

Did 5G Cell Phone Radiation Cause Covid-19?

7

We live in an era of conspiracy theories, and few are more outlandish than the idea that 5G cell phone radiation caused Covid-19 [1]. One claim is that 5G radiation weakens our immune system so we are more vulnerable to Covid. Another, even more absurd, assertion is that the Covid-19 pandemic is a coverup of 5G health hazards. The Covid/5G conspiracy theory has circulated on Facebook [2] and Twitter [3]. In Europe, the United Kingdom, and Australia, this misinformation has encouraged people to set dozens of 5G cell phone towers on fire. Proponents often point out that Covid-19 originated in the Chinese city of Wuhan, a leader in rolling out 5G technology. This book will not address the coverup nonsense, but instead will focus on the suggestion that the radiation itself has dire biological consequences. Could 5G radiation harm our health?

5G

First, what is 5G? 5G is the fifth generation of cellular phone networking that began deploying in 2019 (Fig. 7.1). It commonly uses frequencies up to 6 GHz, although some proposed technology could range between 30 and 300 GHz [4]. This compares to 4G (fourth generation) cell phone frequencies of about 1 GHz. The higher frequencies of 5G radiation are attenuated in the air more than 4G radiation due to their greater absorption by oxygen and water vapor. Therefore, more lower-power cell phone base stations (towers) are needed. Compared to 4G networks, 5G should have higher download speeds (more bits per second), lower error rates, shorter delays between sending data and receiving a response, and greater connectivity (more devices can join a network). In general, the most powerful cell phone radiation a person is exposed to is from the uplink signals from a phone to a tower, as opposed to the downlink signal from a tower to a phone. Adaptive power control technology, which we discussed in the last chapter, should reduce the uplink signal strength compared to older 4G cell phone signals.

B. J. Roth, *Are Electromagnetic Fields Making Me Ill?*,
https://doi.org/10.1007/978-3-030-98774-9_7

Fig. 7.1 A 5G cell phone tower in Darmstadt, Germany. (IToms, CC BY-SA 4.0, https://creative-commons.org/licenses/by-sa/4.0, via Wikimedia Commons)

Much of what we learned in the previous chapter about cell phones applies to 5G too. However, a few differences exist based on the physics of electromagnetic waves.

Skin Depth

When electromagnetic radiation propagates into a conductor, it decays over a distance called the skin depth (frequently referred to as the penetration depth). The term "skin depth" has nothing to do with the skin that covers our bodies. Instead, it merely refers to a layer at the surface of a conductor that is exposed to the radiation. A copper wire has a skin depth and so does the salt water of the ocean. Because our bodies are mainly salt water, our tissue also has a skin depth (Fig. 7.2).

What causes the skin depth? Recall from the previous chapter that an electromagnetic wave is a dance between electric and magnetic fields; the changing magnetic field creates the electric field, and the changing electric field produces the magnetic field. The electric field makes current flow in the conductor. This current arises from moving charges, such as sodium and chloride ions, or from the rotation of polar water molecules. The current generates its own magnetic field that partially cancels the original field, which causes the electromagnetic wave to decay. The wave is restricted to a layer a few skin depths thick. The skin depth depends on the

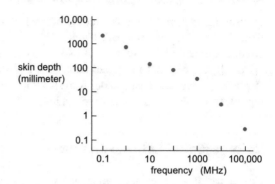

Fig. 7.2 The skin depth versus frequency for muscle. (Data from Adair [5])

frequency of the wave, the conductivity of the tissue, and the rotational properties of water molecules. The fall off with depth is not abrupt; some of the field remains several skin depths in, but, as a rule of thumb, the electric and magnetic fields fall off by about a factor of three for every skin depth of penetration.

At low frequencies, the skin depth in tissue is larger than the size of our body, meaning that attenuation can be ignored. An FM radio station operating at 0.1 GHz (100 MHz, or station 100 on the FM dial) has a skin depth of about 80 millimeters, which is a little more than 3 inches [5]. A 1 GHz cell phone signal has a skin depth of 30 millimeters (about an inch). By 10 GHz, the skin depth is only about 3 millimeters. These waves cannot penetrate far into the body. Deep into the 5G frequency band, at 100 GHz, the skin depth is on the order of 0.3 millimeters, meaning that the wave will literally be absorbed by our skin.

Electromagnetic waves carry energy. Where does this energy go when the wave cannot penetrate into the tissue? Some of the energy is absorbed as heat, produced by either the electrical current created near the tissue surface (Joule heating) or the rotation of water molecules and the resulting frictional heating. At high frequencies, some of the energy is reflected from the body surface rather than being absorbed (your body acts as a mirror).

What are the implications of the skin depth for health hazards associated with 5G radiation? If a 5G cell phone operates at 30 GHz, its radiation is attenuated significantly by the time it passes through the scalp and skull and reaches the brain. The superficial regions of the brain's temporal lobe would still receive some radiation, but not much, and deeper structures would be shielded nearly completely. Ironically, in some ways this makes 5G radiation safer than 4G; the shorter skin depth for 5G means most of the brain is shielded and receives a lower dose. Most of the incident energy is reflected or absorbed near the surface of the body, putting the skin and the cornea of the eye at highest risk.

The idea that 5G radiation cannot penetrate deeply into tissue has been questioned by Martin Pall, of Washington State University, who believes that the electric field in an electromagnetic wave is unable to penetrate into tissue, but the associated magnetic field can. In a recent article, he stated that "electrical fields are almost completely absorbed in the outer 1 mm of the body... however, the magnetic fields are very highly penetrating" [6]. The problem with this idea is that high-frequency electromagnetic waves are created by an interplay between electric and magnetic fields; the two are intertwined. If one decays, so must the other. In a letter to the editor commenting on Pall's article, Kenneth Foster and Quirino Balzano write that "[Pall's] premise, that the penetration depth of magnetic fields from an incident [high-frequency electromagnetic] wave is far greater than that for the electric fields, represents an elementary misunderstanding of the nature of electromagnetism" [7].

Voltage-Gated Calcium Channels

University of Pennsylvania engineer Kenneth Foster has called Martin Pall "the most visible scientist in the public arena" [8] on the subject of 5G health risks. Pall is one of the promotors of the Covid/5G conspiracy theory and wrote in an online article that "5G radiation is greatly stimulating the coronavirus (COVID-19) pandemic and therefore, an important public health measure would be to shut down the 5G antennae" [9]. Before the 5G controversy arose, Pall had sparred with Foster and John Moulder about the safety of WiFi (wireless) networks [10–12].

Pall received his undergraduate degree in physics, and then received a PhD in biochemistry from Caltech. He is now retired from Washington State University after a career studying issues such as chronic fatigue syndrome. He has published an online book titled *5G: Great Risk for EU, US and International Health!* [13]

According to Pall, the main mechanism underlying health consequences of electromagnetic radiation is activation of voltage-gated calcium ion channels [13]. Recall that ion channels are proteins in the cell membrane having a water filled pore that allow ions to pass from one side of the membrane to the other. Generally, these ion channels are selective (they allow, for instance, calcium ions to pass, but not sodium or potassium ions) and are gated (they open and close depending on the voltage across the cell membrane, or some other factor). If calcium ion channels opened in response to a radio frequency signal, the influx of calcium into the cell would have a dramatic effect. Calcium acts as a "second messenger" in the cell, controlling signals sent from the membrane to the nucleus, and triggering muscle contraction. In a healthy cell, the intracellular calcium concentration is usually very low. Too much calcium inside a cell kills it.

Pall argues that the forces on the calcium channel are over seven million times larger than the forces on a single ion, which is why these channels are so sensitive to electromagnetic fields. He writes

> Our current safety guidelines are based only on heating (thermal) effects. Heating is produced predominantly by forces on singly charged groups in the aqueous phases of the cell but the forces on the voltage sensor are approximately 7.2 million times higher. Therefore, our current safety guidelines are allowing us to be exposed to EMFs that are approximately 7.2 million times too strong. [13]

Pall identifies three factors underlying this magnification:

1. *"The 20 charges on the voltage sensor make the forces on voltage sensor 20 times higher than the forces on a single charge"* [13]. An ion channel does have

many charges, but it is a large membrane protein. Its mass is much greater than that of a single ion, so you need a larger force to move it or make it open.

2. *"Because these charges are within the lipid bilayer section of the membrane where the dielectric constant is about 1/120th of the dielectric constant of the aqueous parts of the cell, the law of physics called Coulomb's law, predicts that the forces will be approximately 120 times higher than the forces on charges in the aqueous parts of the cell"* [13]. Pall is right that the dielectric properties of water do help shield ions in water from electric fields, and this shielding should not be as effective in fat, which makes up the membrane. The dielectric constant of water, however, decreases dramatically at high frequencies [5]. In addition, the charges on an ion channel are usually located near the water interface instead of being embedded deep in the fat, so Pall may have exaggerated this effect.

3. *"Because the plasma membrane has a high electrical resistance whereas the aqueous parts of the cell are highly conductive, the electrical gradient across the plasma membrane is estimated to be concentrated about 3000-fold"* [13]. This concentration happens at low frequencies, but at high frequencies the membrane capacitance provides a short circuit, eliminating this effect.

Taken together, Pall's factor of seven million shrinks considerably.

Even if the factor of seven million were correct, several theoretical considerations limit how well a calcium channel could respond to a radio frequency field. We learned in the last chapter that the time required for one oscillation of a high-frequency wave is much shorter than the time it would take for a single calcium ion to cross the cell membrane; for 10 GHz radiation, one oscillation is a thousand times shorter. Moreover, a massive membrane protein could not respond on the time scale that the electric field is oscillating; big and bulky ion channels cannot open and close at gigahertz frequencies. Instead, a nonlinear rectification process would be needed to demodulate a pulsed radio frequency signal, and nobody has ever figured out how such a mechanism would work [14].

Perhaps scientists have not been smart enough to uncover such a mechanism. What is the experimental evidence that calcium channels are responsible for the biological consequences of 5G cell phones? In 2021, Andrew Wood and Ken Karipidis, of the Swinburne University of Technology in Australia, reviewed 50 years of publications examining calcium channels and electromagnetic fields, with a focus on assessing Pall's claim that calcium channels are exceptionally sensitive to radio frequency electromagnetic radiation. They noted that results vary, even about the direction of the response; some articles report that microwave radiation increases the calcium concentration inside the cell, and others report radiation decreases it. They cited several studies that have searched for the rectification and demodulation behaviors that would be necessary for any plausible mechanism to work, and failed to find them. They could find no reason to expect that calcium channels are more susceptible to microwave radiation than other channels having similar structures, such as channels for sodium and potassium ions. They concluded that "experimental studies have not validated that [radio frequency radiation] affects [calcium ion] transport into or out of cells" [15]. Despite Pall's insistence otherwise, the data argue against voltage-gated calcium channels playing a significant role in the 5G safety debate.

Joseph Mercola has recently published the book *EMF*D*, in which he compared using a cell phone to smoking. He wrote that "they are each an enormous threat to individual and public health" [16]. Who does Mercola cite to support this claim? He cites Martin Pall and his calcium channel hypothesis. Stephen Barrett of Quackwatch, a website (https://quackwatch.org) maintained by the Center for Inquiry, said that Mercola is the "world's most dangerous supplier of health misinformation" [17]. Mercola spreads wrong information about other health issues besides just 5G cell phone radiation. The Center for Countering Digital Hate placed Mercola at the top of its list of the "Disinformation Dozen" responsible for two-thirds of the anti-vaccine information about Covid-19 [18].

Foster and Moulder have critically reviewed scores of experiments examining the biological consequences of radio frequency waves on fetal development, the immune system, the elevation of stress markers, and other topics [10]. As usual, the data are mixed, but they find no consistent evidence demonstrating a health risk associated with radio frequency electromagnetic waves. When we add these considerations to the data presented in the last chapter, we find no reason to anticipate that the radiation associated with 5G cell phones should be harmful.

A more recent review by Myrtill Simkó and Mats-Olof Mattsson looked specifically at studies using frequencies corresponding to 5G. The majority of the articles they examined reported seeing biological responses associated with electromagnetic radiation. They assessed these articles using several quality criteria (adequate control studies, blinded studies to avoid bias, data from a range of radiation intensities to develop a dose-response curve, etc.), and found most studies failed to meet them. They identified temperature control as a prime factor in these experiments; if temperature is not rigorously held constant, warming produces artifacts in experiments. Their verdict: "the available studies do not provide adequate and sufficient information for a meaningful safety assessment, or for the question about non-thermal effects" [19].

Mattsson, Simkó, and Foster formulated a list of guiding principles for high quality radiofrequency electromagnetic field studies, with the goal of improving the quality of the research. They wrote:

> Such measures are expensive and time consuming for investigators. But it is inexcusable to encourage more poor quality studies that can alarm the public, and at the same time fail to meet quality standards for inclusion in health agency reviews and well-done systematic reviews. The present need is for well done, adequately supported studies using standard protocols, particularly at 5G high band frequencies.[20]

As Foster and Moulder have noted, "impossibility arguments are difficult to sustain in biology" [10]. Like them, I can offer no proof that 5G radiation is absolutely safe. But the weight of evidence is against 5G radiation causing cancer. I would say solidly against it. I expressed this opinion at a town hall meeting in Rochester, Michigan while explaining the health hazards of 5G radiation, but the audience did not agree with me (Fig. 7.3).

Fig. 7.3 I discussed the health hazards of 5G radiation at a town hall meeting in Rochester, Michigan in 2019. The audience was not convinced by my claim that the risks are small

Déjà Vu

The 5G cell phone debate strikes me as déjà vu. First Mesmer's "animal magnetism" treatments ascended in popularity and then declined. Next the use of magnets for therapy rose and fell. Then came the power line debate; a crescendo followed by a diminuendo. Later the dispute over traditional cell phones came and went. Now, we are doing it all over again for 5G.

The IEEE Committee on Man and Radiation has summed up the potential health risks of 5G radiation this way: "COMAR concludes that while we acknowledge gaps in the scientific literature, particularly for exposures at millimeter wave [5G] frequencies, the likelihood of yet unknown health hazards at exposure levels within current exposure limits is considered to be very low, if they exist at all" [4].

Should We Trust Science?

A disturbing aspect of the debate over 5G cell phone radiation is the tendency to mistrust science. The conspiracy theory about a link between 5G radiation and Covid-19 is the most flagrant example of this. A relationship between Covid-19 and 5G radiation is rejected by the biomedical community. For such an association to be hidden by "elitist experts" would require a cover-up involving thousands (perhaps tens of thousands) of scientists—physicists, biologists, chemists, epidemiologists, virologists, medical doctors—from every country in the world.

A large fraction of Pall's online book dealt with the alleged corruption of federal agencies in the United States and their refusal to take his radio frequency health claims seriously. He wrote that the Federal Communications Commission has acted "in wanton disregard for our health" [13]. The subtitle of Devra Davis's book *Disconnect*, discussed in Chap. 6, included *The Truth about Cell Phone Radiation, What the Industry Has Done to Hide It* [21]. Paul Brodeur's book *Currents of Death* (see Chap. 4) asserted there is an "attempt to cover up [the power line field] threat to your health" [22]. Robert Becker, in *The Body Electric* (Chap. 4), declared that "lay people… cannot automatically accept scientists' pronouncements at face value, for too often they're self-serving and misleading" and that "science is becoming our enemy instead of our friend" [23].

Becker, Brodeur, Davis, and Pall tell stories about how scientists have lost funding for their research investigating how electric and magnetic fields affect our body and attribute these loses to corruption. I wonder if the real reason might be because funding agencies concluded that research on radio frequency hazards was unlikely to lead to important results. Might those agencies have decided that their scarce funds ought to be spent on more promising studies?

Many times I have participated in reviewing grant proposals for the National Institutes of Health and the National Science Foundation (Fig. 7.4). I have never seen any evidence of deception. Instead, these agencies bend over backwards to avoid even the slightest conflict of interest during a review. They ensure each proposal receives the fairest hearing possible. I have, however, witnessed reviewers agonizing over difficult and painful decisions about how to score proposals based on the research's significance and innovation.

I don't believe there's a cover-up. Sometimes proposals are not funded because scientists do not think the research will advance the field. When a reporter for the *New York Times* asked Eleanor Adair about the calls for more spending on research into health effects of electromagnetic fields, she responded that the money "could be much better spent on other health problems. Because there is really nothing there" [24].

Is there a conspiracy to hide the danger of cell phones? If so, all the biomedical researchers who conducted the critical reviews and meta-analyses that I have cited would had to be complicit in the scheme. Funding decisions on grants and agency recommendations are not made by bureaucrats; they are made by consensus among

Fig. 7.4 Peer reviewers evaluating grant proposals for the National Institutes of Health. (From https://public.csr.nih.gov/ForApplicants/InitialReviewResultsAndAppeals/InsidersGuide)

experts, our nation's top scientists. No one is trying to keep anything secret. Scientists don't hide; they seek.

This is not to say that scientists never make mistakes. The reason for peer review is to force scientists to convince other scientists that their ideas and data are sound. It is a self-correcting process, which occasionally makes short-term errors, but finds the truth in the long run. Scientists who challenge existing doctrines are valuable, because they force researchers to reassess their fundamental assumptions and beliefs. On rare occasions, a maverick scientist overturns the prevailing dogma with a novel paradigm shift. Far more common, however, is that when mavericks find themselves opposed to ideas widely-accepted within the scientific community, it means the mavericks are wrong.

Dangers arising from cell phone radiation strike me as unlikely, but not inconceivable. However, the claims that there exists a vast plot, with scientists colluding to conceal the facts, are ridiculous. Scientific consensus arises when a diverse group of scientists openly scrutinizes claims and critically evaluates evidence [25]. Science is the best way to debunk conspiracy theories and uncover the truth.

References

1. Meese, J., Frith, J., & Wilken, R. (2020). COVID-19, 5G conspiracies and infrastructural futures. *Media International Australia, 177*, 30–46.
2. Bruns, A., Harrington, S., & Hurcombe, E. (2020). "Corona? 5G? or both?": The dynamics of COVID-19/5G conspiracy theories on Facebook. *Media International Australia, 177*, 12–29.
3. Ahmed, W., Vidal-Alaball, J., Downing, J., & López Seguí, F. (2020). COVID-19 and the 5G conspiracy theory: Social network analysis of Twitter data. *Journal of Medical Internet Research, 22*, e19458.
4. Bushberg, J. T., Chou, C. K., Foster, K. R., Kavet, R., Maxson, D. P., Tell, R. A., & Ziskin, M. C. (2020). IEEE Committee on Man and Radiation—COMAR Technical Information Statement: Health and safety issues concerning exposure of the general public to electromagnetic energy from 5G wireless communications networks. *Health Physics, 119*, 236–246.
5. Adair, E. R. (2002). Biological effects of radio-frequency/microwave radiation. *IEEE Transactions on Microwave Theory and Techniques, 50*, 953–962.
6. Pall, M. L. (2021). Millimeter (MM) wave and microwave frequency radiation produce deeply penetrating effects: The biology and the physics. *Reviews on Environmental Health*.
7. Foster, K. R., & Balzano, Q. (2021). Comments on Martin Pall, "Millimeter (MM) wave and microwave frequency radiation produce deeply penetrating effects: The biology and the physics," Rev Environ Health, Published online May 26, 2021. *Reviews on Environmental Health*.
8. https://blogs.scientificamerican.com/observations/5g-is-coming-how-worried-should-we-be-about-the-health-risks. Access date: January 12, 2022.
9. https://www.electrosmogprevention.org/5gcovid-19-m-l-pall-3-30-2020. Access date: January 12, 2022.
10. Foster, K. R., & Moulder, J. E. (2013). Wi-Fi and health: Review of current status of research. *Health Physics, 105*, 561–575.
11. Pall, M. L. (2018). Wi-Fi is an important threat to human health. *Environmental Research, 164*, 405–416.
12. Foster, K. R., & Moulder, J. E. (2019). Response to Pall, "Wi-Fi is an important threat to human health". *Environmental Research, 168*, 445–447.
13. Pall, M. L. (2018) *5G: Great Risk for EU, U.S. and International Health! Compelling Evidence for Eight Distinct Types of Great Harm Caused by Electromagnetic Field (EMF) Exposures*

and the Mechanism that Causes Them. https://www.bibliotecapleyades.net/archivos_pdf/5g-emf-hazards.pdf. Access date: January 12, 2022.

14. Foster, K. R., & Repacholi, M. H. (2004). Biological effects of radiofrequency fields: Does modulation matter? *Radiation Research, 162*, 219–225.
15. Wood, A., & Karipidis, K. (2021). Radiofrequency fields and calcium movements into and out of cells. *Radiation Research, 195*, 101–113.
16. Mercola, J. (2020). *EMF*D 5G, Wi-Fi & cell phones: Hidden harms and how to protect yourself*. Hay House.
17. https://quackwatch.org/11ind/mercola. Access date: January 12, 2022.
18. https://www.counterhate.com/disinformationdozen. Access date: January 12, 2022.
19. Simkó, M., & Mattsson, M.-O. (2019). 5G wireless communication and health effects: A pragmatic review based on available studies regarding 6 to 100 GHz. *International Journal of Environmental Research and Public Health, 16*, 3406.
20. Mattsson, M.-O., Simkó, M., & Foster, K. R. (2021). 5G new radio requires the best possible risk assessment studies: Perspective and recommended guidelines. *Frontiers in Communications and Networks, 2*, 724772.
21. Davis, D. (2010). *Disconnect: The truth about cell phone radiation, what the industry has done to hide it, and how to protect your family*. Dutton.
22. Brodeur, P. (1989). *Currents of death: Power lines, computer terminals, and the attempt to cover up their threat to your health*. Simon & Schuster.
23. Becker, R., & Selden, G. (1985). *The body electric: Electromagnetism and the foundation of life*. William Morrow.
24. Kolata, G. (2001, January 16). A conversation with: Eleanor R. Adair; Tuning in to the microwave frequency. *New York Times*.
25. Oreskes, N. (2019). *Why trust science?* Princeton University Press.

Did Cuba Attack America with Microwaves?

8

In 2016, staff in the US and Canadian embassies in Havana, Cuba reported medical symptoms including ringing in the ears, fatigue, and dizziness (Fig. 8.1). The condition suffered by these officials is now known as the Havana syndrome. It continues to affect people and remains a mystery to this day.

Although patients suffering from the Havana syndrome were originally restricted to Cuba, cases have now been diagnosed all over the world, including Guangzhou, China; Bogota, Columbia; Vienna, Austria; and Berlin, Germany. A case was even reported in Washington, DC, where a National Security Council official reportedly became ill near the Ellipse, the oval lawn south of the White House. In August 2021 Vice President Kamala Harris's diplomatic trip to Vietnam was delayed for several hours due to a possible Havana syndrome case in Hanoi. In September, a member of CIA director William Burns's team reported symptoms during a trip to India. And then in October, President Biden signed the Helping American Victims Afflicted by Neurological Attacks (HAVANA) Act, providing support for those suffering from the condition.

What is the cause of the Havana syndrome? No one is certain. One hypothesis is that it results from an attack on American diplomats, perhaps perpetrated by Cuba or Russia. For instance, in October 2017 former President Trump said in a Rose Garden press conference that "I do believe Cuba's responsible. I do believe that." However, we have scant evidence supporting this conclusion. Initial speculation centered on acoustic weapons based on infrasound (sound at frequencies too low to hear) or ultrasound (sound at frequencies too high to hear). A Canadian investigation suspected pesticides were the culprit (Cuba had been fumigating to kill mosquitoes that were spreading the Zika virus). In this chapter, we review what is known and not known about the Havana syndrome and focus, in particular, on the possibility that directed microwave weapons are responsible for the condition.

© The Author(s), under exclusive license to Springer Nature Switzerland AG 2022
B. J. Roth, *Are Electromagnetic Fields Making Me Ill?*,
https://doi.org/10.1007/978-3-030-98774-9_8

Fig. 8.1 Hotel Nacional, Havana, Cuba, the site of a Havana syndrome incident. (Jongleur100, Public domain, via Wikimedia Commons)

The Medical Evidence

Several articles in medical journals have been published about the Havana syndrome. A 2018 article by researchers from the University of Pennsylvania School of Medicine that was published in the *Journal of the American Medical Association*—arguably the preeminent medical journal in the United States—concluded that:

> persistent cognitive [memory problems, mental fog, unable to concentrate], vestibular [poor balance, dizziness, nausea], and oculomotor [eyestrain, light sensitivity, difficulty reading] dysfunction, as well as sleep impairment and headaches, were observed among [21] US government personnel in Havana, Cuba, associated with reports of directional audible [sounds, ear pain, ringing in the ears] and/or sensory [pressure and vibration] phenomena of unclear origin. These individuals appeared to have sustained injury to widespread brain networks [like for a concussion] without an associated history of head trauma. [1]

The paper was controversial. In a commentary about it in the journal *Cortex*, the journal's editor, Sergio Della Sala, and his colleague, Roberto Cubelli, noted that the group from Penn defined "impaired" as performing below the 40th percentile on any clinical test, which is an unusually high threshold. Moreover, when performing more than one test to explore a particular function, impairment was defined as falling below the 40th percentile on any one of the several tests. Della Sala and Cubelli concluded that there is no actual evidence of any cognitive deficit. They went so far as to say that "the condition suffered by the US diplomats in Cuba has been labelled 'mysterious'… The real mystery though is how such a poor neuropsychological report could have passed the scrutiny of expert reviewers in a first class outlet" [2]. Rarely in the polite world of science do you see such sharp criticism of another researcher's article.

Another publication, from a group at the University of Miami and published in the journal *Laryngoscope Investigative Otolaryngology*, found those suffering from the Havana syndrome had damage to their inner ear, resulting in dizziness and difficulty balancing [3]. Yet, only 2 of the 25 patients examined performed abnormally on a hearing test, and both of them had preexisting hearing loss prior to being stationed in Cuba. The authors compared the victims to others living in the same household who did not suffer any symptoms.

A second paper by the University of Pennsylvania group performed more detailed magnetic resonance imaging (including functional MRI and diffusion tensor imaging), and compared victims to normal controls (the original article did not compare their findings to a control group) [4]. The authors found small but statistically significant differences in brain volume and other brain properties between the two groups, although they were not able to determine if these differences were clinically important.

The Frey Effect

Most of those suffering from the Havana syndrome reported hearing sounds when their symptoms began. These sounds, which appeared to originate from a particular direction and to be perceived at a specific location, are the reason some officials initially suspected a sonic attack and are key pieces of evidence pointing toward a microwave weapon.

The microwave auditory effect, also known as the Frey effect, is the perception of audible sounds caused by modulated microwave radiation. The phenomenon was first reported in 1962 by American neuroscientist Allan Frey [5]. In 1974, Kenneth Foster and Edward Finch were able to reproduce the Frey effect in water [6]. They found that the behavior was produced by the absorption of intense 2.5 GHz microwave energy. The resulting temperature increase is tiny (less than a thousandth of a degree Celsius) because the microwave pulse is so brief (a few microseconds). Although small, this heating makes the tissue expand slightly, which launches a pressure wave that propagates outward, often conducted by bones. The inner ear (the cochlea, not the part of the inner ear responsible for balance) then hears this sound as a soft click.

The Frey effect is an example of a nonlinear, demodulation phenomenon, like those discussed in the previous chapters. Heating depends on the square of the electric field, which has two implications. First, if you double the electric field strength, you increase the heating by a factor of four. Second, and more importantly, the heating does not depend on the sign, or direction, of the electric field. Heating by an electric field in one direction is not undone when the electric field reverses. Therefore, a rapidly oscillating electromagnetic wave produces a net source of heating. Moreover, if the amplitude of the oscillating wave is varied, or modulated, at a slower frequency, the heating varies with the modulation. This is just the sort of mechanism needed for a biological response associated with a radio frequency wave, especially one you hear as sound.

Microwave hearing takes advantage of the exquisite sensitivity of the ear. The pressure wave produced in Foster and Finch's experiment is minute. The only reason you are able to hear it is because evolution gave us the ability to detect faint sounds. Such a pressure wave is no more likely to your damage the brain than any other slight sounds you might hear. You don't get a concussion listening to whispers.

In order to scale up the microwave intensity to create sounds so loud they would damage tissue, James Lin, of the University of Illinois at Chicago, has estimated that every square meter of the wave would need to contain 15 GW of power [7]. Fifteen gigawatts is seven times the output of the power station at Hoover Dam. Although the power would be emitted in brief pulses so the total energy and temperature rise would not be so dramatic, and although the beam might have a small cross-sectional area reducing the total energy emitted, the equipment needed to generate such a high-power microwave beam would still be substantial. Foster dismisses the possibility of a microwave weapon, saying "the idea that someone could beam huge amounts of microwave energy at people and not have it be obvious defies credibility" [8].

In the next chapter, we discuss the wavelength of an electromagnetic wave. The wavelength limits how well you can focus a beam of radiation. For instance, suppose a Cuban agent was trying to focus a 3 GHz microwave beam, which has a wavelength of about a tenth of a meter. Focusing that beam to an extent much less than a wavelength would be difficult. You cannot make a pencil-thin microwave beam. So, if you need 15 GW of power per square meter, and your beam is limited to having a cross sectional area of at least a tenth of a meter on each side, you need an energy source that emits about 150 MW. That's a lot of energy; sufficient to power over 10,000 homes. Perhaps you could use shorter wavelength radiation and focus it into a narrower beam, but then it would not penetrate into the tissue as far.

The JASON Report

In 2018, JASON—an independent group of elite scientists who advise the government on matters related to science and technology—conducted an investigation into the Havana syndrome. JASON was created in 1960, soon after the Sputnik launch, to answer questions related to defense. The name refers to Jason of the Argonauts, who quested for the Golden Fleece in Greek mythology. In the past, JASON has advised on subjects such as global warming, acid rain, health informatics, cyberwarfare, and renewable energy. The group includes eminent scientists from many fields, but especially from physics, and several members have been Nobel Prize winners. In September 2021, the news organization Buzzfeed obtained a highly redacted copy of JASON's Havana syndrome report through the Freedom of Information Act [9].

In their report, JASON examined cell phone audio and video recordings from affected individuals, including at least one of which coincided with the onset of symptoms. The sounds were loud, but not so loud that they were painful; the victims did not cover their ears to lessen the volume. The sound did not seem to be localized to a beam, but emanated from a specific location. In one case, this location was a back kitchen door; opening and closing the door changed the loudness of the sound. The sound recorded on the phone was reported to be similar to that heard in person. Electronics, such as WiFi and TV reception, were not disrupted when the sound was present; you would expect such interference from an intense microwave beam [10].

In their report, JASON reached three conclusions:

1. The recorded audio signal is, with high confidence, not produced by the nonlinear detection of high power radiofrequency or ultrasound pulses. [10]
2. We judge as highly unlikely the notion that pulsed RF [radio frequency] mimics acoustic signals in both the brain (via the Frey effect) and in electronics (through RF interference pickup). [10]
3. We believe the recorded sounds are mechanical or biological in origin, rather than electronic. The most likely source is the Indies short-tailed cricket, *Anurogryllus celerinictus*. [10]

After examining the frequency content of the recorded sounds, the report concluded that the call of these crickets matched, "in nuanced detail," the recordings (Fig. 8.2) [10].

JASON's second conclusion was crucial, because if the sound arises from the Frey effect, which occurs inside a person's head, that means it is not associated with a sound wave in the air. Cell phones would not record such sounds unless it was through electronic interference. It is highly improbable that such interference would sound exactly like the noise heard in person by those afflicted by the Havana syndrome.

Fig. 8.2 An adult male Indies short-tailed cricket *Anurogryllus celerinictus*. (Fig. 1a from Sheyla Yong (2019) *Ecologica Montenegrina*, 20:163–167. CC BY)

The National Academies Report

Owing to the unresolved questions surrounding the cause of the Havana syndrome, the US Department of State asked the National Academies of Science, Engineering, and Medicine to review the evidence and make recommendations [11]. The National Academies is a non-profit organization founded by Congress in 1863, charged with providing independent, objective advice to the nation on matters related to science and technology. The Stevens Report about power line magnetic fields and cancer, which we discussed in Chap. 4, was prepared by the National Academies.

A 19-person committee was charged with preparing the National Academies report. It was chaired by David Relman, a microbiologist from Stanford University. The report, published in 2020, considered four possible explanations for the symptoms: chemical exposure, infectious disease, psychological issues, and directed pulsed radio frequency energy. The committee found no evidence for chemicals, such as insecticides, or for infectious agents such as the Zika virus. Those causes would affect not only the embassy diplomats, but also the local staff and residents living near or working at the embassy. Furthermore, it contended that the sudden onset of strange symptoms was not consistent with a psychological factor. Because the sounds that victims heard seemed to be associated with a particular direction and location, the committee suggested that radio frequency energy was the most plausible explanation for the symptoms [11]. The National Academies committee did not seem to be aware of the JASON report, which was completed 2 years earlier but was classified so it was not readily available.

The report found that people did not feel any warming, nor was there any thermal damage evident in medical images. It therefore excluded intense radio frequency energy that is associated with heating. The report next turned its attention to nonthermal mechanisms. At this point, it cited individual articles supporting nonthermal biological responses to radio frequency radiation. We have already seen that the literature is so enormous and inconsistent that you can make arguments either for or against the biological effects of radio frequency radiation depending on which references you choose. The report did not cite many review articles or meta-analyses that critically assessed much of this research, nor did it address the lack of biophysical mechanisms.

Instead, the report cited reviews of radiotherapy, despite its use of x-rays and other ionizing radiation, which is completely unlike the nonionizing radiation from microwaves. It cited a discussion of microwave therapy applied through a catheter to destroy tissue, despite that being an obviously thermal result. It cited microwave studies at very high frequencies without noting that the skin depth would prevent these waves from penetrating into the brain. It cited the blood brain barrier work by Salford, without mentioning that these results were inconsistent with other studies (see Chap. 6). It cited a review of electromagnetic hypersensitivity, a phenomenon in which a person claims to be affected by weak electromagnetic fields, even though medical research suggests that electromagnetic hypersensitivity is not real [12]. It cited the coherent vibrations postulated by Herbert Fröhlich, without mentioning the theoretical limitations of such a mechanism (described in Chap. 6). Finally, when discussing voltage-gated calcium channels, it cited two papers by Martin Pall, the researcher who promotes a relationship between 5G radiation and Covid-19.

Next the report considered the Frey effect, suggesting that it could explain why the victims of the Havana syndrome heard noises. It claimed that a "Frey-like" effect might induce responses similar to those created during transcranial magnetic stimulation to treat depression. Magnetic stimulation, however, is based on a well-understood mechanism: a strong (1 T) rapidly changing (0.1 millisecond) magnetic field induces an electric field in the brain that is sufficient to excite neurons (see Chap. 5). The report does not explain why there should be any connection between the Frey effect and transcranial magnetic stimulation, except that they both involve electromagnetic fields.

In defense of the National Academies report, the committee had limited evidence to draw on, and the report did stress the tentative nature of its conclusions. The clinical symptoms displayed by the Havana syndrome victims were heterogeneous and nonspecific. In addition, due to privacy issues the committee did not have access to individual health records, so it had to base its deliberations on aggregate data. Nevertheless, it took a less skeptical view of the scientific literature surrounding the biological consequences of radio frequency electromagnetic radiation than do many of those who have reviewed this subject, including myself.

The National Academies report pointed a finger at microwave weapons, which the JASON report had already ruled out. Instead, JASON concluded that the sounds were produced by crickets, but chirping crickets could not cause the neurological damage reported by the clinical studies at Penn and Miami. If not microwaves or crickets, then what? Is there any other plausible explanation for the Havana syndrome? One: psychogenic illness.

Mass Psychogenic Illness

In their book *Havana Syndrome*, Robert Baloh and Robert Bartholomew suggest that the US diplomats reporting these peculiar symptoms are suffering from a mass psychogenic illness [8]. This condition, also known as mass hysteria, arises when symptoms spread without any physiological cause. However, the National Academies committee had dismissed a psychogenic origin of the illness, saying that the abrupt onset of unusual symptoms argued against it.

Baloh and Bartholomew present a lengthy list of examples of psychogenic illnesses [8]. They note that diplomats exhibited similar symptoms as those suffering from war trauma, including post-traumatic stress disorder. Psychogenic illness has long been associated with mysterious sounds. For example, the "wind turbine syndrome" refers to people who report becoming sick from the noise of turning wind turbine blades. A psychogenic illness is thought to be an example of a nocebo. While a placebo produces beneficial effects based on positive expectations, a nocebo produces harmful effects based on negative expectations. Baloh and Bartholomew also describe outbreaks of mysterious illnesses, some with neurological symptoms, associated with hostile foreign governments. The diplomats and their families knew they were being watched constantly by Cuban officials. Stress and fear frequently generate clusters of unexplained psychogenic symptoms. The embassy staff were counseled about the threat of a mysterious illness, which may have contributed to its spread (an example of confirmation bias, where patients report symptoms they were told to expect). Psychogenic illness can be accompanied by physiological changes in the brain triggered by chronic stress. Baloh and Bartholomew describe the situation in Havana as "a classic set-up for an outbreak of mass psychogenic illness" [8].

Baloh and Bartholomew stress that just because an illness is psychogenic does not mean that a patient's pain and suffering is phony [8]. The reason for a psychogenic illness is not that the patient is weak-minded or irrational. Certainly, nobody believes that the victims are making up stories or "faking it." Psychogenic illness affects normal people; it is not restricted to those suffering from mental illness. Attributing the Havana syndrome to a psychogenic origin is *not* disrespectful to the victims, just as a diagnosis of post-traumatic stress disorder is not disrespectful to a traumatized person. A psychogenic illness is as real as a concussion; only the cause is different.

The Answer

At this time, we have no conclusive explanation for the Havana syndrome. We need more evidence. Measuring intense beams of microwaves should be easy to do and would not be prohibitively expensive. Until someone observes microwaves associated with the onset of this illness, I will remain skeptical of the National Academies' conclusion.

The National Academies committee did not consider sonic weapons, but never said why not. A sonic weapon would have limitations based on the physics of acoustics, but a microwave weapon also is limited by the physics of electromagnetism. It is not evident to me that a sonic weapon is any more implausible than a microwave weapon.

After examining the evidence, we can make two claims with confidence:

1. If the sounds heard by victims of the Havana syndrome are causing neurological damage via the Frey effect, then the microwave beams must be intense. Their presence would not be subtle. They should be easy to detect.
2. If weak electromagnetic waves are causing the Havana syndrome, then our understanding of the laws of electromagnetism and thermodynamics must be reconsidered. You cannot say that electromagnetic waves are responsible for the Havana syndrome but then rule out their role for causing damage from cell phone radiation. The two problems are inextricably tied.

Baloh and Bartholomew believe that microwave weapons are not responsible for the Havana syndrome. Instead, they maintain the condition is psychogenic. They concluded:

The media, the government and one of the world's most respected medical journals were all pointing in the same direction: US Embassy staff were the victims of a mysterious new weapon that had produced a never-before-seen syndrome. *There is only one problem with this story – it never happened.* There was no secret weapon, and there was no attack. At first glance, the claims appear to have merit, but dig deeper and look at the science behind these assertions, and it makes little sense. The claims are pseudoscience: a set of theories and claims that appear to be grounded in science and facts but do not follow the scientific method. [8]

References

1. Swanson, R. L., Hampton, S., Green-McKenzie, J., Diaz-Arrastia, R., Grady, M. S., Verma, R., Biester, R., Duda, D., Wolf, R. L., & Smith, D. H. (2018). Neurological manifestations among US government personnel reporting directional audible and sensory phenomena in Havana, Cuba. *Journal of the American Medical Association, 319*, 1125–1133.
2. Della Sala, S., & Cubelli, R. (2018). Alleged "sonic attack" supported by poor neuropsychology. *Cortex, 103*, 387–388.
3. Hoffer, M. E., Levin, B. E., Snapp, H., Buskirk, J., & Balaban, C. (2018). Acute findings in an acquired neurosensory dysfunction. *Laryngoscope Investigative Otolaryngology, 4*, 124–131.
4. Verma, R., Swanson, R. L., Parker, D., Ismail, A. A. O., Shinohara, R. T., Alappatt, J. A., Doshi, J., Davatzikos, C., Gallaway, M., Duda, D., Isaac Chen, H., Kim, J. J., Gur, R. C., Wolf, R. L., Grady, M. S., Hampton, S., Diaz-Arrastia, R., & Smith, D. H. (2019). Neuroimaging findings in US government personnel with possible exposure to directional phenomena in Havana, Cuba. *Journal of the American Medical Association, 322*, 336–347.
5. Frey, A. H. (1962). Human auditory system response to modulated electromagnetic energy. *Journal of Applied Physiology, 17*, 689–692.
6. Foster, K. R., & Finch, E. D. (1974). Microwave hearing: Evidence for thermoacoustic auditory stimulation by pulsed microwaves. *Science, 185*, 256–258.
7. Lin, J. C. (2021). Sonic health attacks by pulsed microwaves in Havana revisited. *IEEE Microwave Magazine, 22*, 71–73.
8. Baloh, R. W., & Bartholomew, R. E. (2020). *Havana syndrome: Mass psychogenic illness and the real story behind the embassy mystery and hysteria.* Springer.
9. https://www.buzzfeednews.com/article/danvergano/havana-syndrome-jason-crickets. Access date: January 12, 2022.
10. https://www.documentcloud.org/documents/21068770-jason-report-2018-havana-syndrome. Access date: January 12, 2022.
11. National Academies of Sciences, Engineering, and Medicine. (2020). *An assessment of illness in U.S. government employees and their families at overseas embassies.* The National Academies Press.
12. Rubin, G. J., Hillert, L., Nieto-Hernandez, R., van Rongen, E., & Oftedal, G. (2011). Do people with idiopathic environmental intolerance attributed to electromagnetic fields display physiological effects when exposed to electromagnetic fields? A systematic review of provocation studies. *Bioelectromagnetics, 32*, 593–609.

Is That Airport Security Scanner Dangerous?

9

In an article posted on the *Scientific American* blog on December 18, 2017, physician Farah Naz Khan asked "is that airport security scanner really safe?" She was talking about the advanced imaging technology scanners, in which you stand in a booth that rotates around you, irradiating you with microwaves (Fig. 9.1). Khan writes "until there is proof that the machines… are 100 percent safe even with long-term chronic exposure, I will continue to opt-out of [advanced imaging technology] screenings" [1].

Science can never guarantee that anything is "100% safe." Everything in life carries some risk. But let us assume her 100 percent safe comment is rhetorical. Should we be concerned about the safety of advanced imaging technology? Are there any risks we are ignoring?

Wavelength

Last chapter we briefly introduced the concept of wavelength. Let us examine that idea in more detail.

Another name for advanced imaging technology is "millimeter wave scanners." The name comes from the wavelength of the electromagnetic radiation the scanners use. The wavelength is the distance between two adjacent peaks of a wave. Recall that we know from Maxwell's equations that electromagnetic waves travel at the speed of light, and they can oscillate at a variety of frequencies. The wavelength (meters per oscillation) is simply the speed (in meters per second) divided by the frequency (oscillations per second, or hertz). A 60 Hz power line wave has a wavelength of 5000 kilometers, which is nearly as big as the radius of the earth. An AM radio station emits a signal that has a wavelength of about 300 meters, roughly a fifth of a mile. Your microwave oven uses radiation with a wavelength of 0.12 meters, which is 12 centimeters, or about 5 inches. For a frequency of 30 GHz (used by airport scanners) the wavelength is down to 1 centimeter, or 10 millimeters (four

© The Author(s), under exclusive license to Springer Nature Switzerland AG 2022
B. J. Roth, *Are Electromagnetic Fields Making Me Ill?*,
https://doi.org/10.1007/978-3-030-98774-9_9

Fig. 9.1 A passenger in an advanced imaging technology scanner at an airport. (From: https://www.pnnl.gov/nationalsecurity/millimeterwave/learn_more.stm)

tenths of an inch). Electromagnetic waves with wavelengths of a few millimeters are called "millimeter waves."

What is the significance of the wavelength? Last chapter we saw that the wavelength limits how well you are able to focus a wave. If you are doing imaging, the wavelength sets a limit to the resolution of the image, through a process known as diffraction. For instance, a millimeter wave scanner produces an image of your body, but getting a resolution of better than a few millimeters is difficult. A few millimeters is sufficient to detect a knife or gun, but it is not sufficient to obtain a detailed image of it. Visible light has a wavelength of about half a micron (0.0005 millimeters), which is why light microscopes can take precise pictures of cells (which are typically about 10 microns in size). However, a light microscope will visualize a virus (usually about a tenth of a micron) as a fuzzy blob, at best.

If half a wavelength matches the size of an object, you can get a resonance where absorption is particularly strong (the body acts as an antenna). Whole body absorption of microwaves has a broad peak in the range of 30–100 MHz, which corresponds to a wavelength of 3–10 meters in air. The wavelength is smaller inside the body because light propagates more slowly there (tissue has a high dielectric constant). So half a wavelength in tissue is roughly the same as the size of the body in this frequency range, causing a resonance response. Such an enhanced response occurs only if the skin depth of the radiation is larger than, or at least similar to, the wavelength.

Terahertz Radiation

A frequency of 1000 GHz is called a terahertz (THz) [2]. Terahertz radiation (sometimes referred to as "T-waves") reaches up into the infrared part of the electromagnetic spectrum, and corresponds to frequencies of approximately 300–30,000 GHz (in other words, wavelengths from 1 to 0.01 millimeters). A wavelength of a few hundredths of a millimeter means that high resolution imaging is possible with terahertz radiation; perhaps not at the cellular level, but with higher resolution than microwaves provide.

Terahertz radiation cannot ionize atoms. The photon energy for terahertz frequencies ranges from about 0.001 to 0.1 electron volts, which is too weak to break all but the most tenuous of chemical bonds. However, these sorts of energies are similar to the rotational, vibrational, and translational energies of many molecules. Therefore, T-waves may be useful for doing spectroscopy: examining tissue properties over a range of frequencies to probe how the different molecules, radicals, and ions of a tissue rotate, vibrate and translate [2].

As discussed in the previous chapter, high frequency electromagnetic radiation is not able to penetrate deeply into tissue. At 1 THz, the penetration depth is less than a tenth of a millimeter, so all the energy is reflected or absorbed in the skin. Terahertz radiation might be useful for imaging skin cancer or for colon cancer using an endoscope similar to that employed during a colonoscopy, but it is useless for imaging tumors located several millimeters or more below the tissue surface. T-waves can penetrate insulators that do not contain much water (clothing, paint, cardboard, and sometimes even walls), offering the intriguing prospect of seeing through objects that are opaque to visible light (like superman's x-ray vision, but without the ionizing x-rays). Terahertz radiation is absorbed by the atmosphere more than microwave radiation used for 5G communication. Using T-waves for a sixth generation of cell phone technology (6G) will be challenging.

Airport Scanners

Airport scanners operate at about 30 GHz, which is below the terahertz range and corresponds to a wavelength of 10 millimeters and a penetration depth of a millimeter. The waves pass through clothing but reflect off the body. The reflected wave is what the scanner detects to create an image. By irradiating the body with radiation emitted at many angles, and then using a computer to process the data, the operator can quickly obtain an image of the body surface. The time for a scan is less than one and a half seconds. The radiation is present for only about a third of that time, so you end up with about half a second of exposure each time you pass through airport security, usually once per flight.

Because of potential health concerns such as those mentioned by Farah Naz Khan, the Department of Homeland Security, which operates the Transportation Security Administration (TSA) responsible for airport safety, asked the National Academies to prepare a report about advanced imaging technology used in airports. The committee was chaired by Kathryn Logan, an engineer at the Georgia Institute of Technology. Its report, "Airport Passenger Screening Using Millimeter Wave Machines: Compliance with Guidelines," was published in 2017 [3].

The report noted that originally two different technologies were adopted to screen passengers: millimeter wave machines and x-ray backscatter systems. However, in 2013 the use of x-ray backscattering was discontinued, in part because of the public's fear of ionizing radiation, even though the x-ray dose was so low that it is comparable to the dose received from cosmic rays during a few minutes of flight. You can read about the pros and cons of x-ray backscattering in a *Medical Physics* point/counterpoint debate of the proposition "backscatter x-ray machines at airports are safe" [4].

The National Academies committee then analyzed the intensity of the millimeter wave radiation that passengers were exposed to. Its report quoted the radiation intensity in units of power per unit area (watts per square meter, or W/m^2) rather than specific absorption rate. The generally accepted safe maximum intensity to avoid tissue heating is 10 W/m^2. The airport scanners use an intensity of 0.0001 W/m^2, which is less than the safety threshold by a factor of 100,000.

To translate this into specific absorption rate, assume all the incident energy was absorbed in the first 1 millimeter of tissue, and that the density of the tissue is similar to that of water (1000 kilograms per cubic meter). The specific absorption rate becomes 0.0001 W/kg, which is over 10,000 times smaller than the 1.6 W/kg threshold for safe exposure set by the Federal Communications Commission. If part of the energy is reflected, the amount absorbed is even less. Moreover, the only tissue that would receive this dose would be that within a millimeter of the body surface, and it would be exposed to this radiation for less than 1 second.

The report went on to note that the cornea of the eye is at particular risk not only because it represents part of the body surface, but also because the tissue is poorly perfused with blood so it cannot dissipate heat as effectively as skin can. Nevertheless, the safety factor of more than 10,000 between the allowed specific absorption rate and the rate arising from the scanner should mean that even the eye is adequately protected from any thermal injury.

The National Academies report did not examine nonthermal mechanisms of biological effects, claiming the committee was charged with determining if airport scanners exceeded the established intensity threshold to avoid heating, not with determining if that threshold is adequate to ensure safety.

What do we conclude from the National Academies report? As a result of the short time of exposure, the weak intensity of the radiation, and the lack of penetration, millimeter wave scanners should be harmless. The relative risk is less than the risk from talking on your cell phone for a minute, and we already found that the risk of cell phones is small. Khan may continue to opt out of millimeter wave screening when she flies, but there is no reason you should. As we have noted before, impossibility arguments in biology are…well…impossible, but the risk of any health hazard from an airport scanner is so remote that you could probably think of hundreds of things you do every day that carry greater risk. I suspect the risk from opting for a pat down search during airport screening (oops, did my elbow smack you in the eye when I leaned over…so sorry) is far greater than the risk of a millimeter wave scan.

References

1. https://blogs.scientificamerican.com/observations/is-that-airport-security-scanner-really-safe. Access date: January 12, 2022.
2. Smye, S. W., Chamberlain, J. M., Fitzgerald, A. J., & Berry, E. (2001). The interaction between Terahertz radiation and biological tissue. *Physics in Medicine and Biology, 46*, R101–R112.
3. National Academies of Sciences, Engineering, and Medicine. (2017). *Airport passenger screening using millimeter wave machines: Compliance with guidelines*. The National Academies Press.
4. Hindié, E., Brenner, D. J., & Orton, C. G. (2012). Backscatter x-ray machines at airports are safe. *Medical Physics, 39*, 4649–4652.

Conclusion

10

After reviewing so many medical devices, procedures, and risks associated with electromagnetic radiation, I will now go out on a limb and offer my views about which we should worry about and which we should not. Please take these as subjective opinions based on my qualitative grasp of the evidence, not as objective conclusions based on some quantitative assessment like meta-analysis. Below, I list 12 topics in rank order, with the best understood first and the least plausible last. I divide them into three categories: the firmly established, the questionable, and the improbable. For short, you can call these categories "yes," "maybe," and "no."

Firmly Established

1. *Heating.* Microwave ovens do it every day.
2. *Electrical Stimulation of Nerves and Muscles.* Pacemakers, defibrillators, and neural prostheses are miracles of modern medicine. Transcranial magnetic stimulation is an invaluable tool for studying the brain.
3. *Deep Brain Stimulation, Electroconvulsive Therapy, and Rapid Rate Transcranial Magnetic Stimulation to Treat Depression.* All have much support, but are not quite as well-established as pacemakers and neural prostheses.
4. *Magnetoreception and Electroreception.* The evidence is strong, but in some cases the mechanisms are still being debated.

Questionable

5. *Transcutaneous Electrical Nerve Stimulation.* A toss-up. The clinical trials are inconsistent but lean negative. Yet, TENS has a long history of clinical use. Distinguishing a real effect from a placebo effect is difficult.
6. *Bone Healing.* Like TENS, the clinical trials are not too supportive, but the technique has been around a long time. Too close to call.
7. *Transcranial Direct Current Stimulation.* I worry about how small the electric field is in the brain. The placebo effect probably dominates, but the evidence is too mixed to say for sure.

Improbable

8. *Microwave Weapons and the Havana Syndrome.* I am skeptical about microwave weapons, but so little evidence exists that I want to throw up my hands in despair. My guess: the cause is psychogenic. But if anyone detects microwaves during an attack, I will reconsider.
9. *Cell Phone Radiation and Cancer.* I have serious reservations about the alleged health hazards, both for traditional cell phones and for 5G.
10. *Risks from Millimeter Wave Airport Scanners.* The scanners are safe.
11. *Powerlines and Cancer.* Debunked years ago.
12. *Magnets for Pain Relief.* Pseudoscience.

Some of the most striking lessons I have learned while working on this book is that the scientific literature on these topics is enormous, and the results are mixed. You can find dozens of studies that support any point of view. That is why I have relied so heavily on critical reviews and meta-analyses, in order to get the big picture averaged over all the individual studies. But even these reviews and analyses are not unanimous. You cannot avoid making your evaluations based on inconsistent evidence. Nevertheless, in most cases a scientific consensus has emerged.

The study of biological effects of weak electric and magnetic fields attracts pseudoscientists and cranks. Sometimes I have a difficult time separating the charlatans from the mavericks. The mavericks—those holding nonconformist views based on evidence (sometimes a cherry-picked selection of the evidence)—can be useful to science, even if they are wrong. The charlatans—those snake-oil salesmen out to make a quick buck—either fool themselves or fool others into believing silly ideas or conspiracy theories. We should treat the mavericks with respect and let peer review correct their errors. We should treat the charlatans with disdain. I wish for the wisdom to tell them apart.

Should technologies not be adopted for the public until they are proven safe, no matter how small the risk? Or should they be viewed as innocent until proven guilty? Clearly there needs to be some middle ground. If you are waiting for science to give you a guarantee of safety, you will wait forever. Nobody is able to prove beyond a shadow of a doubt that power lines do not cause cancer, magnets fail to reduce pain, or those airport scanners are 100% safe, but to me the evidence supporting their safety is overwhelming. I find the evidence to be convincing that cell phones are not dangerous and that the Havana syndrome resulted from something besides microwave weapons. Yet the one thing I am *most* certain of is that the public debate will continue. Scientists have an obligation to engage in this debate.

Index

Printed in the United States
by Baker & Taylor Publisher Services